MW00463445

WOUNDED TIGRIS

WOUNDED TIGRIS

A RIVER JOURNEY THROUGH
THE CRADLE OF CIVILIZATION

LEON McCARRON

PEGASUS BOOKS

NEW YORK LONDON

WOUNDED TIGRIS

Pegasus Books, Ltd.
148 West 37th Street, 13th Floor
New York, NY 10018

First Pegasus Books cloth edition November 2023

ISBN: 978-1-63936-507-4

10 9 8 7 6 5 4 3 2 1

Printed in the United States of America
Distributed by Simon & Schuster
www.pegasusbooks.com

To those who live on the river, and keep it alive

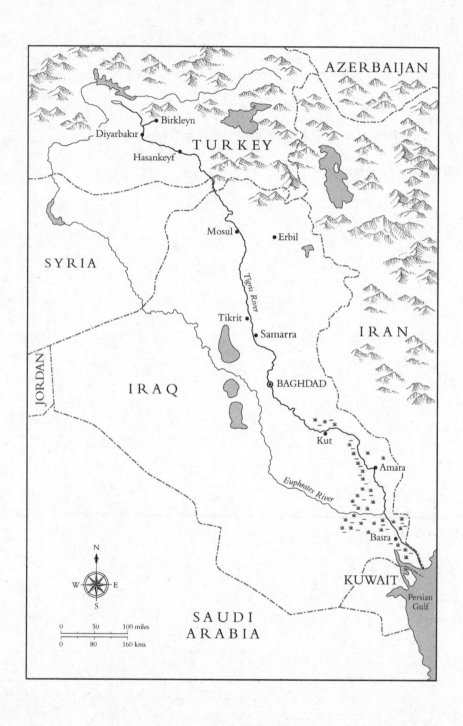

Contents

The country, as you are well aware, is a boiling cauldron. For a foreigner to come and sail in a raft, boat or any other kind of vessel would be absolute suicide. Who would guarantee his safety and wellbeing with all of these criminals and cut-throats having their spheres of influence all over the country? He'd be a sitting duck before he sets his feet on board his boat. Please tell the gentleman that this envisaged adventure should be postponed till better, much better, times come and stay for keeps.

It is of course an insane undertaking, but those are the only ones worth doing.

Even if you are not harassed by ISIS, your life will be made miserable by the local authorities.

The naysayers come from all parts. Keep your pecker up!

Four responses to the concept, pre-journey, each helpful and instructive in their own way. I have given each adviser the benefit of anonymity, and am particularly grateful to the Iraqi friend of a friend who issued the first statement and, in many ways, was not far wrong.

Key Characters

Leon McCarron   Writer and adventurer, working in Iraq since 2016. Co-founder of a long-distance hiking trail in the Kurdistan Region of Iraq.

Emily Garthwaite   British photojournalist and artist based between Akre and Erbil in the Kurdistan region, and passionate about the culture and heritage of Iraq. Involved in the expedition from the outset.

Claudio von Planta   A Swiss filmmaker with decades of experience in Iraq. Best known for filming *Long Way Round* with Ewan McGregor and Charley Boorman.

Angel Istek   Kurdish fixer, journalist and interpreter from south-east Turkey.

Bişar İçli   River expert in Diyarbakır. Also an organic farmer and jack of all trades.

Salman Khairalla   Environmental activist and co-founder of Iraqi NGO *Humat Dijlah*. Spent much of the past decade travelling to places located on Iraq's waterways to train and empower local communities.

Hana Ibrahim	Iraqi translator and activist, with a passion for abandoned animals and environmental campaigning.
Ali Al-Kharki	Environmental activist and co-founder of *Humat Dijlah*.

Prologue: Twin Rivers

The miracle of Mesopotamia begins deep in the great sweep of peaks and lakes that form the eastern Taurus range, beneath an unremarkable mountain called Korha. There, in a narrow, steep-sided canyon, a rough stream tumbles along its base until it reaches a cavernous opening, thirty feet high. The entrance is violent, boulder-strewn. Damp air fills with a rumbling that starts deep inside and exhales against the current as a perpetual roar. There is rarely anyone around to see it but, at dawn, mist like breath drifts out from eddying pools in the unlit mouth of the chasm.

This, the Assyrians decided three millennia ago, was the source of the Tigris. Close by rises the Euphrates and, to the south, an otherwise arid and resource-poor floodplain is transformed. Now we call that area the Cradle of Civilisation, where the birth of the city-state led to the invention of the wheel and the written word. To that list of firsts we could also add codified legal systems, schools, sailing boats, beer brewing and love songs. For those and more, gratitude should be paid to the Tigris, along which I travelled, and on whose current you will join me in this book. But before that, let me tell you what else I know of the river.

For thousands of years the Tigris and Euphrates have acted as a lung, breathing life into what the Greeks called

Mesopotamia, or 'the Land Between Two Rivers'. That name stuck, at least from the end of classical antiquity until a flurry of new terms in the last couple of hundred years culminated with the post-Ottoman breakup. Now most of ancient Mesopotamia is encompassed in the modern state of Iraq, a hundred-year-old guardian of the history of the world.

The two rivers are best understood as twins, born from a single, deep trench. That was during the last Ice Age, sixteen thousand years ago, and when the ice melted the two entities we know now were separated. Their courses have changed a little over the years, particularly in the south where the gradient is gentler, but mostly their nature and direction have remained consistent. From the headwaters the Euphrates meanders around modern Turkey, finding manic courses through high mountains, then pitches east as if heading for the Mediterranean. In Syria it comes within a hundred miles of the sea then relents, turning back towards the Tigris and settling into a broad, slow channel. The Tigris is the keener of the two, shorter but faster. It pierces south through the heart of Iraq and the two meet at Qurna, close to the marshes that historically regulated the flow of the rivers. From there they surge as one to the Gulf, joined by the Karun coming in from Iran.

Water, historically at least, was everywhere in Mesopotamia, from the northern mountains to the southern marshlands, and moved cyclically. Precipitation up high met with snowmelt that drained into tributaries that fed the rivers that washed to the Gulf. This evaporated and was carried back to the north to begin again. When my life first led me to go walking in northern Iraq, which is also a part of this story, I wondered how it would be to follow those showers to the ocean.

The geography of the Taurus and Zagros required inhabitants follow the seasons as nomads and, in the deserts beyond the watershed, Bedouin moved among a network of wells to survive. But in the lowlands the alluvial deposits from the rivers created a vast plain, capable of great abundance but limited by a lack of natural resources. What it needed was for humankind to learn to settle, to divert the rivers into tessellations of irrigation canals, and to work this resulting fertile earth.

Around 10,000 BC, hunter-gatherers came down from the mountains to the north of Mesopotamia, where there was still enough rain to nurture a culture of nascent crop-planting and animal husbandry. These were the first farmers and, forty centuries of practice later, they began slowly to move farther south, onto the floodplain. Within another four millennia settlers had spread throughout what is now modern Iraq, and the formation of early city-states at natural ports and confluence points followed their agricultural mastery. In time, these became major centres of power and influence. Political capital flowed up and downstream with the products of the ancient commercial world. Great cities and cultures grew, but always they depended on the mood of the rivers. The annual inundations created years of plenty and years of scarcity depending on the volume and timing of rainfall and snowmelt. Too much water, or not enough: these were the concerns of the Mesopotamian cultivators who created a civilisation rivalled only by the ancient Egyptians.

Sometimes we have an encounter, or a series of encounters, and it's only afterwards we realise they were not fleeting like the others. Sometimes, if we follow each channel of experience back towards its source, like strands of a braided river, then we may realise that those moments of chance changed our lives. That is how I ended up living in Iraq and dreaming of following the Tigris.

In 2016 I came to the autonomous Kurdistan Region of Iraq for the first time. For six years until that point, since leaving university, I had been almost continuously on the move. I'd spent two of those years on a bicycle, then walked across China and through the Empty Quarter desert. A series of human-powered expeditions followed, and increasingly I focused on the Middle East. Initially I travelled because I was curious about the world, and my place in it, and then eventually because I knew nothing else. Pursuing a career in travel writing made me a professional generalist. I liked to think I was collecting knowledge of the many wondrous ways in which the world works. A friend put it differently. 'You're the most unemployable person I know,' he said. Although I didn't recognise it at the time, I arrived in Kurdistan jaded from constant change, and dubious about my ability to say anything meaningful if I never stayed anywhere long enough to know my way around without a map.

On the same day I landed in Kurdistan, by coincidence, a coalition army began the war to liberate Mosul from ISIS, and my commercial flight from Istanbul took its place on the tarmac in the city of Erbil alongside an array of military aircraft. I was a writer, arriving at a moment of great significance, so I did what I thought I should. I went to the front lines, where Kurdish Peshmerga forces and the Iraqi army and militias engaged ISIS jihadists in the towns and villages on the plains between Mosul and Erbil. It was terrifying, because I had never known war, and the suffering of those I interviewed was unfathomable. I soon retreated back to Erbil, feeling out of my depth and with little to offer as a reporter.

There, chastened, I met a young Kurd called Lawin who took me hiking instead. The mountains were as much a part of the story of this land as the war, he said. It was also much more my 'beat', and was as beautiful as anywhere I'd been.

For the next two years, whenever I had the time and money, I returned again and again to walk more. I had just written a book about modern hiking trails which reimagined the old ways of the Holy Land, so perhaps it was only natural that Lawin and I began mapping Kurdish shepherds' trails. We stitched them together to join with pilgrimage routes and old Peshmerga paths. Soon we had a vision of crossing the region on foot between villages. We called it the Zagros Mountain Trail, and it would eventually grow into a multi-year project to promote sustainable tourism. I took great pride in the potential of the trail to make a positive impact on the region, as well as the fact that I'd eventually walked so far I'd wandered back into employability.

The rest of the time I kept travelling, following stories and hunches, writing for magazines and doing occasional bits of work for TV and radio. But on a boat in the Arctic and in the jungle in Nicaragua, I found myself thinking of Iraq. In 2018, I walked the Arba'een pilgrimage between the holy cities of Najaf and Karbala. Among the chaotic hospitality of fifteen million Shi'a devotees I was introduced to a British photojournalist called Emily Garthwaite. We shared a pleasant but brief evening with mutual friends outside a holy shrine, then went our separate ways. I rented an apartment in London in an attempt to put down roots. Some months later Emily and I met again, over lunch at the Frontline Club, and by the following year, to our mutual surprise, had fallen for one another.

Emily had also been a frequent visitor to Iraq, drawn in by the same fascination and curiosity as me. I admired her work, but more still her ethos. She immersed herself in places so they were never simply stories, and I had seen in Iraq how much she cared, and was cared for. In our relationship we found a space to share both the joy we knew there, and our concerns of trying to understand and portray a place so different from our own.

It was Emily's decision to relocate in the winter of 2019. We had only been together a very short time, and were wary of giving up our freedoms. But her idea of moving to Kurdistan was exciting. It felt bold, for our work and our relationship. I found London unfulfilling, and I didn't like the person I became there. British politics depressed me. A move to northern Iraq, ironically, seemed to offer more stability than staying in the UK. Erbil could be a new start, and I could keep working on the hiking trail. So we made a home there, and walked in the mountains near the Turkish border. I learned Arabic from a Syrian tutor in a Jordanian coffee shop, and practised with Kurdish-speaking Iraqis who thought me even stranger than I really am. Emily photographed communities along the trail, and I wrote my stories from a desk in a small apartment on the sixteenth floor of a new-build high-rise.

We travelled to Baghdad and Basra, and to the remains of the great ziggurat of Ur. I spent a few days in the reeded marshlands, which I was told might be the Garden of Eden. Strung out as they initially seemed, I realised over time and repeated visits how these cities and sites of Iraq were all bound together as products of the Tigris–Euphrates river basin. Everything I saw in the country, north or south, existed because of the rivers, and was fastened to them as surely as a boat to a dock.

It was also clear that the rivers and their tributaries have become clogged, erratic. Upstream countries are building dams to control the flow of the life source. A changing climate has brought environmental instability to countries already burdened with other tensions. Poor governance, corruption and archaic water management systems mean vast quantities of water are wasted, and what is left is contaminated in myriad alarming ways.

In the marshes I saw Edenic waters polluted and poisoned, and the marsh Arabs who lived there struggling to continue their riparian way of life. In Basra, where all the threats to the rivers converge in blistering toxicity, our friend Ameer confided that he sometimes had sleepless nights about what the future held for his five-year-old son, Mo. 'I've brought him into a hell,' he told us. 'How do I justify that to him?'

Inside the country, people talked about the very real danger of the rivers running dry. Internationally, Iraq's water crisis rarely made the news, and almost never beyond the middle pages of the papers. But what would it mean for humanity to lose one of the great rivers of civilisation? And was it inevitable? The idea to follow the Tigris took hold of me. Perhaps if I had lived somewhere else, or with someone else, it might have been different. But I was in Iraq, and Emily wanted to come.

It would be a journey in three parts, from highlands to foothills to flood plain. I imagined a small team with a local expert as a companion. Emily would photograph, I'd write, and we'd bring a filmmaker. We would travel in local boats, from source to sea, to get as close to the river as possible: a frog's-eye view, as the swimming author Roger Deakin might have it. The whole thing seemed difficult but achievable. I should have known better.

In the last months before we set off, the Pope made a historic visit to Iraq. He followed a similar itinerary to my early travels, I thought, though grander, stopping at Ur, Baghdad, and meeting the Grand Ayatollah Sistani in Najaf. Finally, he came to Mosul, and rounded out the tour with mass in the football stadium in Erbil. Tens of thousands of Iraqi Catholics gathered in the stands and, on the pitch, close to a makeshift altar, nuns, politicians and journalists pressed together in an uneasy union of VIPs. Francis arrived in his Popemobile at mid-afternoon and zipped around the running track. Black-clad security guards ran alongside as he waved up to the stands. The nuns sprinted to be close, habits in the air, plimsolls discarded to the wind, and they screamed their joy. I ran too, caught up in it all, though I was no match for the zest and elbows of the nuns.

When he rounded the corner, just a few feet away, I thought he looked genuinely happy. His speeches had been powerful throughout and, even as a secular Irishman, I was impressed. 'How cruel it is that this country, the Cradle of Civilisation, should have been afflicted by so barbarous a blow, with ancient places of worship destroyed and many thousands of people – Muslims, Christians, Yazidis and others – forcibly displaced or killed,' he had said. 'We reaffirm our conviction that fraternity is more durable than fratricide, that hope is more powerful than hatred, that peace more powerful than war.'

To me, Iraq was full of hope. It was a hope that came from what had prospered on the alluvial plain. I saw it every time I went hiking and was invited into the home of a stranger, and on every visit to the south when I met young, energetic activists and idealists. The Pope had lamented the tragedies and celebrated the characteristics of the country I'd come to love, and the media frenzy around him couldn't help but carry that message to the world. For the first time in my life, Iraq, the gift of the twin rivers, was being broadcast to the world for a reason other than conflict. It was the greatest source of encouragement I could have hoped for.

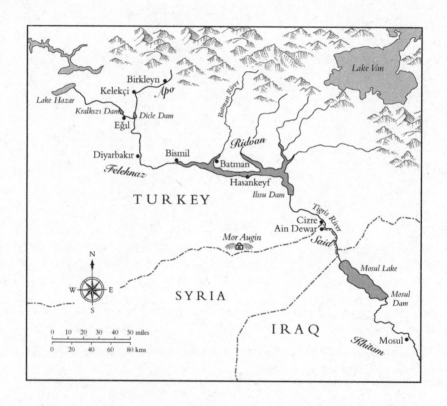

PART ONE

UPPER TIGRIS

TURKEY, SYRIA & KURDISTAN
REGION OF IRAQ

In those ancient days, the lord of broad wisdom, Enki, the master of destinies ... took in his hand waters to encourage and create good seed; he laid out side by side the Tigris and the Euphrates, and caused them to bring water from the mountains; he scoured out the smaller streams, and positioned the other watercourses ... Enki made spacious sheepfolds and cattle-pens, and provided shepherds and herdsmen; he founded cities and settlements throughout the earth.

'Debate Between Bird and Fish', Sumerian essay, 2100 BC

A river flowed out of Eden to water the garden, and there it divided and became four rivers. After the Pishon and Gihon, 'the name of the third river is the Tigris, which flows east of Assyria. And the fourth river is the Euphrates.'

Genesis 2:10–14

The fame of these two rivers, the Tigris and the Euphrates, is such that we need not describe them.

Ibn Jubayr, *Travels*, 1184

Chapter One

Divine Roads of the Earth

Day 1
Birkleyn
River miles: 0

By the time we arrived in the valley of the headwaters, I had been navigating along a theoretical Tigris for nearly two years. The river had taken up so much of my life that on good days I dreamed of floating on a boat, mid-Tigris, watching Iraq drift by. Then there were bad days when Emily and I lay on the floor of our Erbil apartment, pinned by the weight of tasks incomplete. At a certain point the best days, which were very rare indeed, became the days when I didn't think about it at all. We needed buy-in from government and security to allow us to travel; colleagues and advisers along the river to connect us with boatmen and tell us what we didn't know; funding to pay everyone. It was significantly more complex than the largely solo endeavours I'd done in the past. It was Emily's first expedition of any kind.

Inspiration never waned, but enthusiasm was at times submerged by the practicalities of Turkish, Syrian and Iraqi bureaucracy. And yet, really, it was how it should be. Someone said offhand that this might well be the first attempt at a full descent of the river since Ottoman times. To prepare quickly for that would not be right. The labour of the preparation matched the ambition, and had it come together with any semblance of simplicity I might have had reason to distrust the whole thing.

We left on St Patrick's Day, under a gloomy sky. Having received permits to travel in Turkey only one day earlier, Emily and I flew to Diyarbakır and rented a car to drive to the source. On this early part of the journey we would have to alternate between boats and the vehicle. The security situation in the south-east, where the conflict between the Turkish army and Kurdish militants has been ongoing since the late 1970s, meant there were sections of river where significant travel on the water was impossible. The area around the source was one of these. Ironically, perhaps, we expected that river access in Iraq would be much easier.

With us was another foreigner, Claudio the Swiss filmmaker, and two Kurds, Bişar the river expert and Angel the translator. We stopped at a petrol station on the way to collect Ziya, who Bişar said lived in a nearby village and knew the area well. Ziya was sixty-five and wore scuffed dress shoes, a torn suit jacket and had one eye. He'd spent five years in prison in the 1990s, for political reasons. Here, this usually meant aiding the PKK, but Bişar didn't elaborate. He said only that Ziya's wife had died a week ago, and the old man would appreciate the company and a chance to earn some money. Ziya spoke little in the car, smiled even less, and sat uncomfortably on Bişar's lap. Everyone was too polite to tell me I should have rented a bigger car, and instead the criticism hung in the air, audible only in the grunts and sighs of fidgeting passengers. We left the main road and drove down a dirt track, over a brook and into a clearing surrounded by high rock walls. Blackened earth and charred branches showed where a fire had been, but it was hard to say how long had passed since it gave out any warmth.

Beyond, to the north, the brows of the high mountains at Bingöl were still dusted with snow, a reminder that winter was departing but not yet gone. The immediate landscape was bare, hillsides dry and mostly grey, though in patches the first shoots of spring were starting to break the monochrome. The range to the south was green on the lower reaches that stretched out to the plains. This valley was caught between seasons, and us with it, beginning but not yet begun.

Early conversations in our group of strangers were stilted. I was used to travelling alone, or perhaps with one other, and our number concerned me. I trusted that the specialisations of each would make it worth sacrificing intimacy, but we needed time together. A shared experience to bond us. I also felt us as an imposition on this quiet place, so it was with polite uncertainty that almost immediately we had to consider adding to our group. Mahmud had seen us turning onto the track and jogged down from the road to find out what was happening. 'I know everything here,' he said in broken English, learned from two years working at tourist resorts in western Turkey. 'You want to see where the treasure is? What about where the river is born from?'

He led us up a narrow path, leaping and bouncing over rocks and pulling budding leaves from the trees. Ziya followed slowly, brown shoes sliding on smooth stone. Occasionally he'd stop to stroke the hard stubble around his moustache and tut to himself. At Mahmud, or the path, or something else, we couldn't say. For a mile we walked like this, unlikely companions, through meadows of Jerusalem thorn trees whose dried circular seeds hung from wispy stems and rattled in the wind like maracas marking our passing. Then we climbed again. A narrow pathway trodden by goats and sure-footed shepherds reached over the shoulder of the mountain. The growl of moving water was always near, sometimes seeming to come from beneath our feet, then the sky, then from everywhere at once. Clouds gathered and darkened the land.

Where one path ended, another started, sliding down a steep and treacherous bank for three hundred feet. Recent rains had washed away mud steps that must surely have once been pressed in. Now it was a chute to serious injury or death. We all crept to the precipice. Restless Mahmud was banished to the back, and we took it in turns to peer over the side, each holding the shirt of another for safety. I craned my neck and caught first sight of the embryonic river, battling through a puzzle of shattered boulders, then disappearing along with the light. The entrance of the tunnel that swallowed both was vast. Dusted pink crown anemones grew horizontally out of the rock beneath us, bright and vulnerable. I identified much more with their fragility than with the sureness of the river. 'It starts here,' said Mahmud. 'This is the Tigris.' He cheered and slapped me on the back.

I looked to Ziya for confirmation. He nodded, then stroked his chin again. Bişar, for whom a sly grin was always waiting at the corner of his lips, pushed for more information and smiled when it came. 'It begins from the spring there,' Ziya pointed, and we followed his finger to the nearest mountain. 'But if you go over that bigger mountain behind, all the springs there feed the Euphrates instead.' We were at the birthplace of not just one river, but two, and the headwaters of the Euphrates were within hiking distance had we so desired. Each drop of rain that snagged on this range would end up in one or other river by chance of geography. We had fallen into the watershed of the Tigris and now would follow it, through canyon, lake and floodplain, until we too spilled out to the sea.

Locally, the site of the source is called Birkleyn, and the fledging river the Birkleyn Çay. Orientalist explorers called it 'the Tigris Tunnel'. The rain was coming down heavily and looked as if it might turn to snow. Mahmud's white T-shirt had soaked through, clinging to his chest hair like a vacuum pack. I dug in my bag for a down jacket and he wrapped himself in it, shivering, pouting. 'Let's go now. Everything else is boring here. I'm hungry too.'

We backtracked, then took a new trail which dropped to the mouth of another cave, tall and broad enough to fit a double-decker bus. The entire valley, Ziya said, was a single band of limestone piercing through a volcanic range, riddled with spiralling voids in the earth. Eventually, we'd come to where the river exited its underground path through the mountain, but first there was something he wanted us to see.

Carved into limestone at the cave entrance at head height, Ziya traced for us the shape of a figure. It stood tall and straight, dressed in a flowing robe. The right hand was pointing, finger outstretched, not dissimilar to how Ziya had directed us to the source of the Tigris. The relief was of Shalmaneser III, King of Assyria from 859 to 824 BC and long-time campaigner in these lands at the north-east extent of his empire. It appeared to us like a halo, the body brown, fading to yellow at the edges, then to white, finally blending back into grey rock. Beside it, distinctive wedged cuneiform characters told how he had battled his way here, and venerated him for great achievements. For nearly three thousand years this glorification has remained, resisting regional foes, weather, looters and now the modern-day political graffiti that covers much of the rest of the wall.

Shalmaneser III left myriad records of his two journeys to

Birkleyn.* In his annals, he recorded: 'In my fifteenth year of reign I marched against the land of Nairi [modern Lake Urmiya, Iran]. At the sources of the Tigris I cut an image of my royal majesty in the cliffs of the mountain, by the place where the water breaks forth. I wrote thereon the glory of my might, the triumphs of my power.'

One way we could think of the next six hundred miles was as a voyage back through the political and military influence of the Assyrians until we reached their heartland in central Iraq. At the time of his campaigns in these edge-lands, Shalmaneser III was contributing to what would become one of the largest empires the world had ever seen.

After the Assyrians, others wrote about Birkleyn sporadically from a distance. Ptolemy noted its co-ordinates in the second century AD, as did later Mamluk- and Ayyubid-era geographers. But there are few accounts of visitors, which perhaps explains the stillness present even today. Colonial travellers came to know of it in 1862 through John George Taylor, a British diplomat in Diyarbakır. Archives from the Royal Geographical Society in London record that, upon presentation of Taylor's findings, it was noted that another Englishman had been to the site before, but had not spotted anything remarkable. This was used as an example to show the difference between 'travellers who carry their eyes in their pockets, and those who use their eyes for the benefit of mankind'.

It's hard not to spare a thought for that other poor man, with his eyes in his pockets, who overlooked the reliefs and found instead only a nice place for a picnic. I asked Mahmud and Ziya if they knew about any of these travellers. Mahmud

---

* These were inscribed on various artefacts unearthed elsewhere in the kingdom, which almost always seem to have evocative names: the Monolith of Kurkh, the Black Obelisk, the Gates of Balawat.

said no, but he was interested if any of them had found treasure. He'd thought about this a lot, it turned out.

'The finger points to where the gold is,' he said of the gesture of the king, the *ubanu tarasa*, common in Assyrian iconography. Of this he was certain. His grandfather told him stories of people getting lucky by following the finger. But recently successes were few. The most well-known expedition during Mahmud's time was a German group, twenty-eight years ago. He didn't know what they found, but suspected it was good because they had fancy equipment. I said they were probably archaeologists rather than treasure hunters. He shrugged. They took things from the ground and didn't tell anyone, he said, so what did it matter what they called themselves? I asked if he'd ever dug here.

'No, no,' he said, suddenly forceful. 'Never.'

'And is it legal to dig here?' asked Emily.

'No. If they catch you, you'll go to jail directly.' He smiled, suddenly sheepish.

'Until the 1950s and 60s there were some tribes who lived in here,' offered Ziya, suddenly. He used to come here fishing as a boy and remembered them. Some recent studies of the cave suggest human activity at the site since the late Neolithic period, meaning that over eight thousand years of habitation only ended during Ziya's lifetime. The treasure hunters had ruined the place, he said, and the military made it hard even to visit now. This was the first time he'd been in over a decade.

Mahmud was still cold and hungry and pleaded to go back to the car. It was Bişar who finally reminded him he hadn't been invited in the first place. He whispered that he didn't want to go alone in case he missed out on some compensation for his services. Reluctantly he handed back my jacket, and in return we paid him for his guiding more than his company and for his leaving as much as his joining.

Ziya stuck with us and led us finally to where the Tigris emerges from the mile-long tunnel. Now it was calm, pooling in curling eddies at the banks. Another relief stood regally on the rock, high above the waterline. This was Tiglath-pileser, etched in place a quarter of a millennium before Shalmaneser III, and he too performed the *ubanu tarasa*.

There is a theory that the complex of openings at the source are associated with a cuneiform phrase, DINGIR. KASKAL.KUR,* which can be translated as 'divine roads of the earth'. These metaphysical paths are gateways to the underworld, where the membrane that separates a spiritual realm from the physical is particularly thin. The Assyrians acknowledged a sacred power here, washing their steel, offering sacrifices. Their predecessors had different gods, but afforded the same reverence to the space. It seemed fitting that we should also mark the occasion.

Bişar and I returned to the car and collected an armful of wooden planks, a hammer and nails, two balls of string and four partially inflated tractor inner tubes. At the water's edge we topped up the tubes, laid the planks out in crosshatch and bound everything together. It was our first vessel of the expedition. A more accurate description might be to call it a partially buoyant transport pallet, with exposed nails. Its total passenger load was one, and ideally a light one at that, and for the first time the lack of other people in the militarised zone was a blessing, because our hopeless construction did not have an audience. Emily, the most nimble, stepped on and veered out, leaving the jaws of the tunnel behind her. I stripped off

---

* I learned this, and much else about Birkleyn, from a paper written by Ömür Harmanşah called '"Source of the Tigris": Event, place and performance in the Assyrian landscapes of the Early Iron Age'.

and dived in after Emily's raft. The cold struck my skin like electricity, and I flailed and wheeled until my blood flow settled. Imbued with the wisdom of cold water, the worries of starting gave way to the acceptance of voyaging.

Sleet became snow, then rain, then stopped, and afternoon sun cast a sublime amber light. After less than a mile, at a bend in the river, Emily and I climbed back onto the bank. We lit a small fire to dry off and returned to our rental car with the team. Soldiers at a military checkpoint on the way in had registered us, and we were warned not to leave the valley by the river. The meander was the limit of our safe travel. Bişar remarked that had they seen our raft, they would not have felt the need to be so strict.

Fifteen river miles separated us from the point at which we could come back to the water, though the journey by road was much farther. Along the way, military patrols scoured grass verges for IEDs. We would travel in parallel with these tensions all the way into Iraq. Ziya said he'd still love to go fishing at Birkleyn, but maybe now I could understand why it didn't feel so welcoming.

'You asked if I care about the history,' he said. 'How can I, when I can't even come to this place to have a barbecue?'

Chapter Two

You Should've Seen It

Day 2
Kelekçi | Pirxakya
River miles: 37

Angel had worked as a fixer for over a decade, mostly chasing conflict in the region to arrange access for foreign journalists. Like talented people in her position around the world, she lived within a paradox: work was only good when her home was on fire, and when it was peaceful money dried up. It was hard to know what to wish for, other than different career opportunities. As we left the checkpoint outside Birkleyn, she sighed. 'I hate politics.'

She was striking looking, in her mid-thirties like me, with a bob of jet-black hair and red lipstick. She could look stern when needed, but usually a smile unfurled across her face if the right amusement blew her way. Following a river was appealing, even if it sometimes became braided with strands of the politics with which she was jaded. Her son Boran was at home in Diyarbakır, and she sent him pictures from Birkleyn. When he didn't reply quickly, she called to check in. It was just her, she said. Boran's father wasn't around. He was fifteen now, rebellious like she had been, into rap.

When she talked about him she rolled her eyes, as mothers of teenage boys have always done. I thought of my mum, in Northern Ireland, who brought me up alone. She was also young then, and beautiful, and shared her twenties and thirties with me by her side in a way that could have been a hindrance to her youth but which she embraced as a gift. Angel struck me as the same, making her way through life in step with her son, involving him in her life and work.

Our team eased into each other's company, learning about who and what mattered to one another through gentle, investigative conversation. Angel sat on the back seat flanking Emily, with Bişar squeezed in on the other side. At checkpoints she took the questions, knowing when to be clear and firm and when to make gentle jokes with sombre young men who were often hundreds of miles from home. With anyone new, she spoke Turkish, switching to the Kermanji dialect of Kurdish with Bişar, and English with us. Her authority was apparent in the way she listened and made suggestions, and it took only a few hours to realise how fortunate we were to have her on board.

Rural roads wound us through long, lateral valleys in pursuit of the river. The nascent Tigris, still the Birkleyn Çay, flowed due east until steep limestone cliffs forced it into a southward bend. We met it at the turn and walked a small track to where the road ended at a collection of squat concrete houses. It felt good to be on foot.

The village was called Kelekçi, built on a sloped hillside. Above, oak trees fought for space with stubby juniper and svelte pine. Narrow cement paths ran between homes, and animal pens had been fashioned out of bundles of dry branches on the margins. A passing flock of free-roaming chickens with feathered feet stopped to watch us. Three

Zaza* women in bright and colourful ankle-length abayas paid us no heed at all.

Where the ground levelled out, a series of springs were channelled together. At this confluence stood a balding man with thick stubble. His name was Apo. Bişar knew him from a previous visit and he would take us downstream by boat. This was how we would move along the river: in the company of men like Apo from villages on the banks, whom we'd pay by the day for boats and stories.

'You have to use your imagination,' Apo said. 'This used to be such a beautiful traditional village.' It was still, I thought, but he preferred the old, now absent, mud-brick homes. In the past, Kelekçi had been the starting point for an ambitious enterprise. Oak was felled in the forest and the timber carried downriver by horse and donkey to Apo's village, Pirxakya.† There it was lashed together with inflated goatskins for buoyancy. These rafts were called *keleks*, and each measured sixteen square feet and could carry a five-tonne load and a four-person crew. Sometimes a small cabin was added, which made me think of Huckleberry Finn on the Mississippi. The vessels' shallow draught allowed them to travel even when the river was low and, because the current at the headwaters was swift, there was little rowing to be done. The job of the skipper, or *kelekçi*, was mostly to avoid obstacles. This village was named after these men.

Apo pointed to the tumbling stream below us. 'These used to be three watermills here,' he said. As a child he'd come from Diyarbakır with wheat to be ground at the mill, then

---

* Zaza are generally considered Kurds, but trace their lineage to Iran and speak an Indo-European language that bears no relation to Kurdish. The population exists only in the south-west of Turkey, and this was the first of a handful of Zaza villages that we passed through.
† Most villages in this area had two names: the official name in Turkish and the older Kurdish name. Pirxakya is a Kurdish name. Kelekçi is Turkish, and the village was called Zohar by the Zazas who lived there.

return by *kelek* from Pirxakya with walnuts and berries. The journey took two days. Fruits and nuts were sold to city merchants at the docks and other, often larger, *keleks* were loaded for longer journeys south. The rafts from the north, having served their purpose, were broken down and skins loaded onto pack animals to be carried back to Pirxakya. The oak logs were sometimes also sold for a further profit.

Ibn Sa'id Al Maghribi, an Andalusian geographer who also noted the source of the Tigris, wrote about the prevalence of *keleks* on the river in 1267, but they had been well used and documented since long before. I had seen them on the great wall friezes of the Neo-Assyrian Empire, now on display in the British Museum. For nearly three thousand years they were the backbone of trade and transport on the Upper Tigris.

A British travel writer, and spy, E. B. Soane, wrote of a journey by *kelek* on the river in 1914 that: 'Under such conditions, fine weather and a broad river that runs at a steady pace without too many shallows and rapids, there is probably no more pleasant method of travelling than by *kalak*. As it proceeds, the raft turns round and round slowly, giving a view of every side.'

The *kelek* was still a fixture on the Upper Tigris until the 1960s, though by that point its influence had dwindled. Now, Apo said, any *keleks* that we might see wouldn't travel far, and in quality and quantity the rafts themselves were unrecognisable from what he knew in his youth, which in turn were a shadow of their former glory.

We drank long slugs from the spring, wincing at the cold, then piled into Apo's white fishing boat that bobbed up and down by the sandy riverbank. At the third tug of the choke, the ten-horsepower engine coughed black smoke and reluctantly turned over. The Birkleyn Çay here was weak, only thirty feet wide, but as we faltered downstream I noticed we were slowly sinking. Apo presented a cracked yellow jug and encouraged me to get to work. By design or dumb luck, the leaking boat took on water at a slightly slower rate than one full-time bailer could remove it. It was a step up from our raft at Birkleyn, but only just.

'Do you feel that?' shouted Apo over the splutter of the engine. There was nothing, which was his point. 'That's the river dying.'

The villagers near the headwaters had started digging gravel and sand from the river to use for construction, he said. The river was losing its bed. What's a river without a bed, he asked rhetorically? People also illegally syphoned off water to irrigate their fields, further stripping its power. Now the water beneath us had become slack, any kinetic energy relegated to the depths below. The issues upstream were compounded by those ahead, too. We were entering the arm of a dam, said Apo. 'From here on, it's all a lake.'

Perhaps because it felt like the distance to the sky had been lessened, blackbirds, swifts, swallows and woodlarks darted around us in the boat. Their ravines were below us now. We busied ourselves by taking turns to scoop out water and it was rewarding work, because we knew if we didn't the boat would sink. In less than thirty miles, the Tigris had already been cut, diverted and flooded beyond recognition. I had expected to reach a point where the history of the river met that of human-kind's actions, but I had not thought it would be so soon.

After an hour, Apo guided us past two rows of houses that rested by the lapping waves. They were lean concrete cubes with too few windows, built to the same plain design. When the downstream dam was impounded, a new water level was marked at almost exactly the midway point of this village. Those living in the lower homes were offered a small payout and told to leave. The residents above were given no money but informed they could stay. They left in solidarity. Mimicking his rhetoric from earlier, Apo asked: 'What use is half a village?'

By afternoon the horizon broadened as we exited the canyon, floating over what must have once been a large valley. Directly below us was Apo's original village.

'It was even more beautiful than Kelekçi,' he reminded us.

The lake was a mile or more wide from one side to the other, and a last breath of winter blew across its surface as a fierce wind. Beside a gravelly beach, Apo had built a simple two-storey concrete *chaikhana*. These teahouses are still ubiquitous in the wider region, from Afghanistan and Central Asia to Turkey and Iraq, and have their roots in the resting places along the old Silk Roads that unspooled west from China. Now they are almost exclusively inhabited by old men, many of whom sit the whole day through drinking and dozing and opining. Apo had painted his place with thick, vertical white and blue stripes, like a neatly wrapped Christmas present, incongruous with the barren fists of rock around it. There were the beginnings of an orchard outside, which he hoped to nurture when good weather came.

Inside, evening light refracted through thick, uneven windows, showering the teahouse in a kaleidoscopic haze. Old men appeared as the sun slipped away, ambling down the pathway in thick jumpers and tight prayer caps. One carried a piece of flatbread and wordlessly tore off a chunk for me. Apo heated water in a blackened kettle over a buckled iron stove, and soon was serving viscous tea in hourglass cups, pyramids of sugar glittering silver at the bottom, then stirred into constellations. Our group sat at a table in the corner. Apo spoke easily with his clients as he hurried between them, refilling tea, seeming to hold one ongoing conversation with the entire room. All over the tea-drinking Middle East and subcontinent, I thought, a similar scene was playing out in thousands of other places just like this. It reminded me of village pubs in Ireland too; a gathering place for old boys who wanted to get out of the house.

Apo talked of his son, who had wanted to marry a girl, but her family wouldn't allow it. Then he married her anyway in secret, and the girl's family demanded two hundred million Turkish lira, or around fifty thousand US dollars. Apo had sold most of what he had to pay up.

'Everything I do is for him,' he said. Later the son walked in, tall and handsome and a little shy, and the old men laughed and teased him, but they also patted him on the back and clasped his shoulder because he was a romantic, and that was to be respected.

We put up our tents outside the teahouse, watched by the bemused tea-drinkers. One man came to offer help, but poked himself in the eye with a tent pole and quickly went back inside. Above, stars pulled on the fabric of the sky, every bit as plentiful and shimmering as the sugar in the teacups. Before we went to sleep, Apo stood with us by the shores of the lake and said once more: 'You should have seen it.'

Over three hundred feet now separated him from the house where he grew up.

'There was an old caravanserai in there,' he said. 'And Byzantine treasure everywhere!'

Perhaps his *chaikhana* was built to replace an original down below. I tried to imagine it but, in the fading light, on a mountaintop by a bitter lake, I could not. Apo left, following in the wake of the tea-drinkers, and walked home towards his new village a mile away. We were alone, and for a while it was quiet, the only sound the lapping of water and the gentle movement of pebbles disturbed by the waves.

Chapter Three

Holy Water

Day 3
Dicle Dam | Eğil
River miles: 52

The Dicle Dam, which created the reservoir that covered Apo's village, is the meeting point of two branches of the Tigris. One branch was ours, from the east. The other came from Lake Hazar in the west, and the two flow towards each another like a Y-shaped divining rod. Hazar is often considered the primary source, further from the mouth than Birkleyn, but recently the government had cut the stream.* Now there is only a diminished watercourse that flows out through the iron-ore-rich area of Maden, mined in recent years into a bare and cratered wasteland.

Apo had offered to arrange an alternative to his perpetually sinking boat. I presumed a modest upgrade, so it was a surprise to find at the dock a thirty-foot-long, two-tier tourist cruiser covered in pink bunting. On the top deck a tarpaulin protected varnished benches, and five flaccid pink balloons

---

* The Assyrians had been there, too. This whole area provided minerals and materials that the base of their empire lacked, so they traded with the north for iron, silver and copper. These were essential for producing weapons for their campaigning armies, and for the adornment of temples and palaces in their capitals.

bounced on the bow. There had been a wedding onboard and now the captain and his cousin were going home to Eğil, a couple of miles from the dam on the western branch.

Eğil has a long history. Perhaps even longer than most places here. It was originally Assyrian, then occupied and defended as a seat of importance by almost every ruling power in the region. Zülküf, the captain, listed for us Romans, Persians and Ottomans, and could also have mentioned Byzantines and Abbasids, and the Armenians, for whom it was the first capital of their third-century BC Sophene Empire.

'They used to call this the Valley of Roses,' Zülküf said, and brought up a picture on his smartphone. It was a photograph of an original, jaundiced print, taken with an old, low-quality phone camera, and now we peered at it through a cracked screen. I guessed and said it looked like it was probably pretty. The valley had been flooded in 1997. That's what Zülküf had been told, anyway. He was only six then, and by the time indistinct memories of childhood had formed into coherent recollections the lake was already well established.

We cruised at a steady speed, only once having to slow to pass under a low bridge in a narrow canyon. All of us rushed to the top deck to watch the tarp scrape under by the width of a finger. Everyone cheered. Signs of empires past appeared on the final approach to Eğil. First it was a small pyramid carved into the mountain on the west bank, then a series of great cuboid shapes with windows knocked through to a hollow interior. Everything was shaped out of bedrock and, as we drifted, limestone cliffs lost their serrated edges and took the form of a city. The dimensions were off, sharp corners blurred into soft contours, but a bygone glory was clear. Some of what we saw were littoral temples, said Zülküf. Others were fortresses, or private homes for the elite. Many were adaptations of natural caves, but a few were sculpted like ancient high-rises out of promontories.

The modern town itself was high on a plain, hidden from the lake behind the walls and watchtowers of an Assyrian castle from the time of Shalmaneser III. The same king who had journeyed to the source of the Tigris had marked this place, too, and somewhere up there was a stele with another depiction of him, facing east with a sword in one hand and a double-edged axe in the other.

By the docks was a modern promenade, but the water level had dropped from previous years and parts of the old town were appearing again. A cement roof from the 1960s poked out, wraithlike. Probably it had been someone's home. Zülküf edged the boat around it. On a sheer cliff face beyond, diagonal openings revealed an Assyrian staircase that stepped its way, inside the mountain, down to the depths of the lake. Now 130 feet lay under the water.

As the water became shallow, it shone clear and intense. Directly below the boat were rows of tombstones, cut and carved simply in the Islamic way. Before the town was flooded, families had exhumed their ancestors. Zülküf's parents carried the bones of his grandparents to a new site over the ridgeline. The headstones, too heavy to move, were left in place, marking now the death of the town instead. Zülküf's family had been running tours on the lake since 2005, making a business from the project that displaced them. 'Most people who come here as tourists don't even know it's an artificial lake,' he said. Behind us, a teenager on a jet ski made donuts on the open water, the engine and the slap of the boat on the surface reverberating around the theatre of rock.

In town, Emily went to the promenade. For years she had been a street photographer, taking candid shots of everyday life, but here she was surprised when the young man she photographed promptly knelt down and proposed to his girlfriend. The couple assumed she was an official tourist photographer who would try to sell them the picture later on, and were confused when she spoke neither Turkish nor Kermanji. So they invited her for lunch, as so many Kurds do when they meet strangers. Eğil was a special place for them. That's why the man wanted to propose there. 'It's got a magic about it,' he said. Emily promised to send them the picture.

Although an unusual circumstance, this was a typical interaction for Emily. She was not like any photographer I had known before. Emily was interested in people and in their relationship with where they were from. She was very approachable, and happy to approach, and anywhere we went she would strike up conversations. Women, especially, opened up to her quickly. She never seemed to ask questions with an agenda, but with genuine curiosity and desire to learn as much about that person as the window of opportunity allowed. Whether or not she could photograph them was never as important as what they had to say. Her other major projects, photographing the Arba'een pilgrimage, and Bakhtiari nomads in Iran, were all multi-year, multi-visit, multi-media endeavours. On the Tigris, she would often say that this journey was just the start, and that if we were successful on the expedition, it would have to be the start of a long-term commitment to the river.

Bişar knew a path to a lookout, and we climbed breathlessly away from the reservoir until we could see the two arms of the Tigris coming together in greeting. It felt good to be tired, breathing heavily, surrounded by interwoven mountains. Above the forested auburn crags was a kestrel, flitting in and out of view as it carved a wide circle in the sky. When it came close, I could see its fanned plumage, russet and magnificent. Somewhere there were Bonelli's eagles, which made homes in the cliffs, too, and Egyptian vultures; wild horned goats roamed plunging valleys and striped hyenas slunk between caves.

We sat on a rock and looked back down to the dam. It was a broad wall of concrete, marked with the huge letters of the state water company, DSI. The dam's name, Dicle, was the same as the Turkish name for the river. There was also another dam further north on the western branch.* Both were part of the Southeastern Anatolia Project – GAP as the Turkish acronym – which has changed the headwaters of the Tigris and Euphrates for ever.

The desire to harness the power of the two rivers for energy generation was first attributed to Mustafa Kemal Atatürk. By the 1970s, the GAP was a dramatic manifestation of that idea, eventually proposing twenty-two dams and nineteen hydro-electric power plants to irrigate 1.8 million hectares of land and produce 27 billion kWh annually, roughly 25 per cent of Turkey's energy needs. It is one of the largest projects of its type anywhere in the world and will protect the country against increasing regional water shortages, as well as giving it the upper hand over its riparian neighbours. Much of the GAP's development had already taken place by the time we arrived, and as the years and component parts have mounted up, so too has controversy increased.

Alongside its energy creation ambitions, the GAP sought to help farmers move away from reliance on rain-fed irrigation, and increase Turkey's food security. Better living standards were promised to the local population, and a safer future in the wake of decades of the Turkish–PKK conflict. But it was no secret that many Kurds felt the dams were another way

---

* That was Kralkızı, or the King's Daughter, named for a nearby rock formation and from it crept a similar network of fjord-like channels.

for Ankara to assert tighter control on the area and fragment
the Kurdish areas. Villages and towns had been displaced.
Apo had complained that his land had been flooded with the
promise of free and reliable electricity. Now, he said, he still
paid for it and he had no land to generate income.

There was also an environmental concern. The riverine
and canyon ecosystems of the Tigris and Euphrates valleys
are home to vulnerable bird species and other unique biodiversity. Earlier development on the Euphrates, in particular
the Atatürk Dam – the third largest in the world – has led to
a severe reduction in habitat provision. The Euphrates soft-
shell turtle is now no longer found along the river that bears
its name in Turkey. The Tigris has remained more viable
as a safe environment to the fauna of the region, relatively
speaking, though it too is now under threat.

The Tigris before us was stopped in its tracks. A few
modest jets spurted out from the dam's outlet. The great mass
of water had become a stream smaller than the Birkleyn Çay.
What was released would flow to Diyarbakır and south to
Syria and Iraq. Every drop of rainfall and snowmelt was now
counted and controlled. There was trouble there, too, with
Syria and Iraq facing increasingly depleted inflow, and in the
absence of a functioning multilateral water-sharing agreement, political rifts have appeared between the neighbours
over shared resources.

Bişar smiled and rolled a cigarette. 'It doesn't look like a
mighty river, does it?'

We had asked many people over many months as to who
cared the most about the Tigris in Turkey, and, without
exception, they had directed us to Bişar. He was a farmer
and a jack of all trades from Diyarbakır. Emily asked how he
became a focal point for river protection.

'Water should be holy,' he began. 'It is life, and everything

here is related to it. Trade, agriculture, cleaning, everything.
It's our life, and it's being destroyed.' When he was younger,
he'd been to another great river in the west of Turkey,
and saw how well looked after it was. The Tigris deserved
that, he said.

His cigarette was done and he pressed the squat stub onto
the rock, then put it back in his pocket. The dams came with
the promise of benefits, he said: jobs and a better harvest.
But the irrigation never came. The grapes that made this
area famous in Turkey no longer grew. The fish couldn't
make their way upstream any more. All that had made life
sustainable had gone. Bişar shook his head. Above, the kestrel
still drew its circles and in the scattered cliffs underneath us
came busy swifts, to and fro. In the distance, a dog barked
relentlessly.

'Turkey treats its waterways as it likes,' Bişar continued.
There was no consideration for anything downstream,
he said. He told us that for the next forty miles, through
Diyarbakır and Bismil, the Turkish state had downgraded
the Tigris from a river to a stream. 'If it was a river, it would
need to be protected, and there are international regulations,'
he said. But the reduced flow coming from the dams gave the
government scope to change the designation, and on a stream
there are fewer restrictions. That removed legal protection
against mining and building on the banks, said Bişar. In his
eyes, this was one of a number of tools used to manoeuvre
around any responsibility for the wellbeing of the river.

It was getting dark. We had lingered too long above the
dam, so we shouldered our packs and walked over the fields
to the nearest town. The mountains fell away to the plains
and to the city of Diyarbakır. The sun set and we left behind
one of the world's most famous streams.

Chapter Four

Newroz

Days 4–6
Diyarbakır
River miles: 80

One version of the origin story of the Tigris tells that the prophet Daniel was instructed by Allah to trace the route of the river with his staff. He was warned to take good care to avoid those in need, in particular, orphans, widows and the poor. If there's any truth to this, then the erratic steep-walled canyon where the two arms of the Tigris meet must surely have been a collecting point for those in need of compassion.

The river is sketched with eccentric meanders, turning back on itself in hairpin bends, leaving great interlinking horseshoes of rock abandoned like islands. Eventually it spills out onto the great plateau of Diyarbakır, where a mazy patchwork of cultivated flat land props up clenched mountains. On this canvas, the Tigris straightens up considerably. Perhaps once there was an abundance of wealth, parents and husbands.

Like a sapphire ribbon, it rolls into the city. Diyarbakır clings tightly to the water on its west bank, homes and castles and protective walls on steep embankments pressed upon one another. Behind them are modern, identical residential projects, strung out irregularly along major highways like cheap pearls on a chain. It is the largest Kurdish majority city in Turkey, and the de facto capital for the population.

There are between twenty-five and forty million Kurds worldwide. Around fifteen million live in south-east Turkey, with other significant populations spread across the contiguous mountain ranges which run between Syria, Iran and Iraq. They are a minority in each of those countries and often said to be the largest ethnic group in the world without a homeland. There is some uncertainty about exactly where the Kurds first came from, but we know they are descended from Indo-European tribes who settled in these mountains north of Mesopotamia sometime in the second millennium BC. They were then, and remain in some places even today, a nomadic people. As in many such environments, other tribes became amalgamated into Kurdish identity through a shared geography, culture and language.

Since the thirteenth century the ranges in which they live have been called Kurdistan, and yet it has never been a country. Because of the rural and tribal nature of the Kurds, this was not of great concern to them, until the ideal of the nation-state spread east from Europe in the nineteenth century. Under the rule of the Ottoman Empire, a series of sporadic Kurdish revolts rose up against the central government with an aim of some form of independent state. But they lacked then, and still today, a united Kurdish leadership. The uprisings were no match for the control and strategic prowess of the Ottomans, who were familiar with ethnic, religious and nationalist unrest. The attempts failed.

An opportunity came after the First World War with the dissolution of the Ottoman Empire, when the Treaty of Sèvres recommended for the first time the creation of a

Kurdish state. The new Turkish leader, Mustafa Kemal, later Atatürk, became an ally of the Kurdish leaders and recruited them to help define the border of a new Turkish state. The Kurds, for their part, continued to be split, decentralised and without experience in nation-building. This was to prove fatal. In 1923, at Lausanne, the Kemalist government signed a new treaty on behalf of Turks and Kurds, which overrode the suggestions of Sèvres and annexed the Kurdish mountains for Turkey. The moment for a Kurdish nation vanished. What lay ahead was a century of turmoil, the results of which are still seen acutely in cities like Diyarbakır today.

The Tigris brought us into the city through parcelled fields of wheat, barley and cotton. It was an auspicious time to arrive. It was New Year's Eve, *Newroz*, for Kurds, Iranians and others across central Asia. The biggest celebration anywhere in the Kurdish populated areas is traditionally in Diyarbakır. *Newroz* also marks the beginning of spring, and I hoped this meant our days of toiling in the teeth of a winter wind were over.

We met Angel's family, who chastised us for being late. Angel's sister gave her a red sequined dress to change into, and together we walked to Newroz Park in the centre of the city. The sisters began dancing then, and didn't stop for many hours. Emily joined them, too, twirling her expedition shirt in a best attempt to match the flowing dresses.

A heavy police presence funnelled everyone into one main entrance, and Angel flashed the press passes she'd arranged. Inside the atmosphere was Kurdish Glastonbury, with more politics. I later found out the attendance was around half a million. We were still travelling during a pandemic, and until just a week before our arrival in Turkey the country had been under curfew. Now I could see no masks other than our own. 'Perhaps this is part of the Turkish plan,' joked one of Angel's cousins. 'Throw the Kurds together and hope we die of coronavirus.'

In the centre of the park burned a great fire and knots of people, mostly young men, gathered around the smouldering embers. Many were dressed in traditional clothes and more still carried flags and banners. The pyre comprised a column of bricks, thirty feet high, wood packed around the base, and another wall of stone around that to stop people getting too close. The fire symbolised the end of winter and a new season of light. Flames licked around the feet of those standing on the protective wall.

The flags were mostly the purple tree trunk, green leaves and white background of the Kurdish People's Democratic Party, or HDP. They were the third largest party in the country and the most influential of those considered pro-Kurdish.[*] Other peripheral leftist parties were represented too and, amid it all, a few young men had headbands sporting the flag of the sworn enemy of the Turkish state, and US- and EU-designated terrorist organisation, the PKK. I looked around nervously, wondering how they could flaunt these colours so openly.

For many Kurds, like Angel's family, *Newroz* in Diyarbakır was about cultural identity. For others it was an opportunity to galvanise opposition to the government; to show strength. A young student approached me, wanting to make sure I knew the history. The PKK were following in the wake of other movements that had sought to fight against the attacks on Kurdishness, he said. There was one name that he spat out with venom.

'Atatürk started this. He closed the newspapers and moved people out of the homes. He was the first one to attack our

---

[*] A friend had summed up this term for me as meaning, in a political sense, a recognition and promotion of Kurdish identity, culture and political involvement.

language. It was Atatürk. He's a hero over there,' he said, gesturing vaguely to the west, the rest, of Turkey. 'Here, he's why we suffer.'

At the far end of the site an enormous stage book-ended the crowd, with enormous yellow signs covered in bunting that said 'Newroz Piroz Be' – Happy New Year. A local photographer and I fell into step looking for a vantage point. 'It's so depressing,' she said, seeing my notebook filled with scribbled snippets of conversation. Most Kurds were happy to be part of Turkey, she told me – I shouldn't forget that. Even the PKK had given up on separatism and anyway, very few people she knew supported them. Kurds like her just wanted their rights. 'Let us speak Kurdish and have a party for Newroz. Let our politicians stand in a democracy.' She shook her head and disappeared into the crowd.

There has been human settlement at the site of Diyarbakır for ten millennia, and a city has stood since the middle of the second century BC. For much of that time it acted as a natural intersection between landscapes, civilisation and culture. This fusion as a commercial hub gave the city a diverse make-up of Kurds, Turks and even Arabs, and a religious amalgam of Armenians, Assyrians, Chaldeans, Nestorians, Jews and Yazidis alongside the majority Muslims. Today, the notion of a vibrant city in which many faiths and ethnicities co-exist is mostly one that is remembered in the accounts of past travellers and in the edifices of buildings that once housed a now-departed people of the city.

One theme that is constant in writings on Diyarbakır is reference to abundant water sources, mentioned a thousand years ago by the Arab geographer Al Muqadassi in almost exactly the same terms as a pilgrim on his way to Hajj a century later, and a Venetian salesman in the sixteenth century. I liked the idea of a companion from a different era, with whom to compare progress, and settled on the Ottoman traveller Evliya Çelebi. Çelebi was from Istanbul, then Constantinople, born into a privileged family in the early years of the seventeenth century. After a youth spent studying Islam, he became a page at the palace of the sultan. His connections there gave him the foundation for the four decades of travel that followed, during which he criss-crossed the Ottoman Empire and surrounding lands. His book, the *Seyhatname*, or Book of Travel, was six thousand pages long and full of the details that his contemporaries left out.* As well as the history of the pashas he visited, he also included the architecture of their castles, the taste and smell of their food, the names of their slaves and the folk and ghost stories told that rippled through the streets of Ottoman cities.

In 1655 and 1656, Çelebi came to Kurdistan. It was his third visit, and by far the most extensive. Eventually, as part of that same phase of his journey, he would move on to Baghdad, and back along the banks of the Tigris through Tikrit, Mosul, Cizre and Hasankeyf. His description of the journey ahead was as good as any today:

> These streams [the tributaries of the Tigris] join together and the river flows at the foot of Diyarbekir on its eastern side and beneath Fıs Kaya . . . then below Diyarbekir, under the bridge where the rafts dock that go to Baghdad and Basra on the Shatt, flowing toward Hasankeyf and Cezre, watering 100 castles and towns and cities (or Medain) and the entrepôt of Mosul and other regions. By the time it reaches Baghdad, as many as 150 great streams have joined

---

* I have often thought of invoking Çelebi's unflinching longwindedness when negotiating with editors about how severely I need to cut my writing, but have not yet been brave enough.

together. Below Baghdad it is joined by the Diyala and Charka and the Greater Zab and other great rivers. Then above Basra at the promontory of Qurna castle the two brother rivers – the Euphrates and the Shatt [Tigris] – join with a single heart and purpose.

There is also a theory that Çelebi may be the cartographer behind a mysterious seventeenth-century map of the two rivers. Across eight double-folio sheets, attached in strips and measuring over ten feet in length, the map shows the rivers from their source to the Persian Gulf. The river is coloured teal, city silhouettes and ships in spidery black and mountains shaded with yellow and pine green. The headwaters spill in honey-coloured streams from a cluster of domed mountains, and the Tigris soon passes through a huddle of square, red-roofed buildings with a green-domed mosque in the middle. This is Diyarbakır and south, smaller but sketched with different architecture, is the town of Hasankeyf, ancient even to Çelebi. By the time the rivers reach Qurna, they are broad and braided, and intricate settlements, palaces and mosques adorn the banks.

The map surfaced in London in 1980 and was acquired by a member of the Qatari royal family. A Turkish academic called Zekeriya Kurşun has written a convincing argument in favour of the map being the work of Çelebi, based on the physical properties of the map – the style is similar to a map of the Nile attributed to Çelebi, and it matches the time period – and the content, in which the various artistic depictions correspond neatly with what Çelebi focuses on in his *Seyhatname*. I had not seen this map for myself but had come across a digital rendering almost two years before beginning my journey. More than almost anything else, it fired my keenness to go, and to make my way through these enchanted landscapes of the Ottoman inks.

On the first day of the new year, we walked the cobbled streets of Diyarbakır's old city, called Sur. Sur is listed on UNESCO's World Heritage List, but in 2015 a third of the historic area was destroyed in Turkish army attacks on armed Kurds who tried to declare the area as autonomous. A warren of winding alleyways forced us into slender gauntlets, some of which dead-ended in rubble. When a young boy pushing a vegetable cart came past, we all squeezed to the side. Terraces jutted out above, closed except for grates on the walls. These were the 'ears of the street', designed so nosey residents could listen to conversations of passers-by. Behind were large, airy courtyards, often with fountains in the middle, fig trees for shade, and tiers of recessed rooms packed around the square. It felt like it was always dusk in Sur, light blocked by the walls and basalt draining any brightness from the sky.

In one of these courtyards sat a handful of old boys on benches. They were dressed like the elderly Kurdish men we'd met in the countryside, with baggy maroon trousers, neatly buttoned shirts and heavy suit jackets to block the wind. All of them fingered prayer beads in their right hand, counting out each pearl. Left hands were plunged firmly into pockets to keep the cold at bay.

After a while, three figures shuffled to the front. One had a deeply lined, handsome face, like an ageing rock star lost in Sur. He sat in the middle, pushed out his chest and pressed his fingers to his temples. Then he sang, low and guttural, holding certain notes for ten, fifteen seconds, his jaw wobbling up and down with vibrato. The voice was the instrument, pulled and pushed in all directions, resonating deeply as the singer strained at his vocal cords. Half an hour passed, and his eyes were so tightly closed that the skin ruffled at the corners. The men on either side swayed gently, breathing against the winter wind, still carefully counting out their prayers with agile fingers. This was the *dengbêj*, the Kurdish sung-spoken oral storytelling, literally meaning *deng*, the voice, and *bêj*, to tell.

Angel whispered a translation of the first story to me. There was a young man who fell in love and went to the mountains. She rolled her eyes. 'It's always love stories and mountains.' There were lengthy descriptions of the mountains, and of the girl, and of the type of love, and each matched the other; the mountains were tall and majestic, as was the girl, and the love was never-ending and glorious at sunset, much like the peaks and ridge lines.

I found Emily sitting inside another small side room with a woman in a navy headscarf and a quilted beige jacket. She had been watching from a distance. Her name was Feleknaz, and she was a *dengbêj*, too. But she'd run into some trouble with the custodians of the Dengbêj House and could no longer perform.

'We want to hear you,' Emily told Feleknaz. 'Is there another place we can go?'

Ten minutes' walk away, through the maze of old streets, we entered another courtyard, and another square room with wooden sofas, firm red cushions and tall frosted windows. Feleknaz sat with her back straight, hands on knees.

'There are a lot of female *dengbêj*,' she said. Women created it, she told Emily, and they sang at home. Much of it was about pain, and the songs were cathartic. But then the men learned it and forbade women from doing it. She had fallen out with the men at the Dengbêj House because she still recited political stories. She also suspected that some just didn't like having a woman around.

Feleknaz had started as a child. Her first songs were lullabies she heard her mother sing when she was milking the animals. 'I'm fifty-seven now,' she said, with two boys and five girls, 'and I sang to all of them as I rocked them to sleep. Sleep, my dear lamb, I said. And then I told them: You are the one making my heart more patient.'

Two old men walked in quietly, looking at their feet, and sat on the other side of the room. I recognised them from the Dengbêj House, and I wondered if they'd been dispatched to make sure Feleknaz didn't say anything controversial.

'I'll sing something for you from a woman's point of view,' she said. 'There are lots like that, and the men will never sing them. This one is about a girl called Nuri and a boy called Rizgar.'

She cupped her right palm around her ear and began to sing, conducting with her left. Her voice soared, higher and richer than the men we'd heard. It echoed around the walls, amplified by more than lungs. There were intricate moments, which her fingers picked out like gentle strings in an orchestra, then booming refrains. The songs the men had sung were visceral, and the sound vibrated through the basalt floor into my bones, but they were not beautiful. This was sublime. The boy, Rizgar, should not have loved Nuri, but he pursued her anyway. In the end he was killed for it, and Nuri would have been too but escaped. To live, when he died for love, destroyed her. Please kill me, she sang to his dead body in her dreams. Take me, not him.

When Feleknaz finished, there was silence, broken only when the old men in the corner said quietly, '*Saad hosh.*' Very good. They had simply come to listen, delighted by the opportunity to hear Feleknaz perform.

Chapter Five

The Business of *Keleks*

Day 7
Hevsel | Fishkaya | Bismil | Batman River confluence
River miles: 137

We spent a night in one of the old homes in which a thousand years of history had been scrubbed until it shone. It belonged to a collective, friends of Angel's, and they also shared a vineyard outside the city. In a chamber under the master bedroom were five hundred bottles of unmarked wine, so we made a roaring fire under the mulberry tree in the courtyard and drank. Bişar sang silly songs, and Claudio told stories about interviewing Osama bin Laden and being put in prison in Pakistan. I had admired Claudio's work for years, and still found it a little strange that now he was filming our journey. The *Long Way Round* series had been one of the things that encouraged me to set out to see the world at twenty-two. Claudio was dedicated to his craft, and never separated from his camera, and he knew a lot about the Kurds. But despite his experience he spent much more time listening than talking, and would happily ask questions for hours, nodding and furrowing his brow as he took it in. The wine released us from the worries of river travel, and in a quiet moment Emily said we'd need more of these evenings if we were to make it with our sanity intact.

In the morning, we moved slower than usual. All of us carried one backpack each so we were mobile and could hike if needed, or jump easily between boats. There was one additional reinforced case, the size of a briefcase, that carried a spare camera. Other than that, we were nimble and well packed. We did laundry in the sink at the renovated home, and, wherever we were, Emily and Claudio made time in the evenings to transfer data from cameras to hard drives. I rewrote notes from a pocket notebook into a larger one and photographed each page as a backup. It remained to be seen how sustainable this routine would be with the demands of constant travel.

Opposite Diyarbakır's fortress is a sweep of green, cultivated land, over a thousand acres in size, called Hevsel Garden. It is on the UNESCO World Heritage List alongside the city walls. The land is farmed but with a wild edge, split into irregular, overflowing allotments of spinach, lettuce, radishes and chard, and dotted with poplar and fruit trees. Diyarbakır, so long seen as the link between mountains and Fertile Crescent, here shows where the two spaces were fastened, bound by the ribbon of the Tigris.

In some accounts, Hevsel was said to be the place where Adam and Eve met after they were expelled from Paradise. Evliya Çelebi wrote at length of the gardens and of the centrality of the river to the Ottoman Diyarbakır:

Diyarbekir's basil gardens and regularly laid out vegetable plots on the bank of the Tigris have no equal in Rum or the Arab lands or Iran. When, in the spring season, the flood period of the Tigris has passed and its limpid waters begin to flow [again] in a stable current, all Diyarbekir's inhabitants, rich and poor alike, move with their entire families to the bank of the Tigris . . .

For a full seven months a merry tumult, with music and friendly talk, is so going on night and day here on the bank of the river Tigris . . . In short, the people of Diyarbekir arouse the envy of the whole world because of the pleasures and enjoyments that they have on the bank of the Tigris.

Nothing had changed, said Bişar. We walked beyond the gardens to a less salubrious part of the river, sunken beneath modern, sharp city edges, where buildings and roads fell away like debris in a mess of dust and disorder. Heavy rain dropped into our eyes when we looked up, and we clambered over a barricade to reach two square concrete buildings pitched behind a row of poplar trees.

This, said Bişar, was Fishkaya, that Çelebi referred to when he wrote of the 'bridge where the rafts dock that go to Baghdad and Basra'. From the other side of the Tigris, a blue *kelek* came towards us through the mist. It was iron, with a felt covering to support two standing men. Oil drums wrapped in blue tarps kept it afloat. A hole had been made in the iron two-thirds of the way back and the pilot punted with a fifteen-foot pole.

The larger of the two men hopped off, his hand swallowing mine in greeting, and he led us to a building, small and dark, but dry. In the corner, another man in a polo neck watched four teapots resting on an open fire. The walls were lined with shisha pipes and discarded cigarette packs. It was where fishermen came to hide.

The big man's name was Zayyat Kelekçier, and Bişar burst out laughing in surprise. 'Kelekçier?' he repeated, incredulous. Zayyat nodded.

'My family ran the first and best *kelek* business in Diyarbakır. My great-grandfather alone had forty *keleks*.' The business extended a full sixty miles north in its prime. 'All the way to a village called Kelekçi,' said Zayyat.

Under the Ottomans, the Tigris and Euphrates were controlled in their entirety, for the first time, by a single imperial power. Faisal Husain, who has written a history of the two rivers during the empire, calls them 'the conduits through which Ottoman power flowed'. Istanbul harnessed the power of these arteries from the fifteenth century onwards, to transport, irrigate and protect. The rivers were reimagined as corridors which connected to the very borders of Europe. Their major riparian cities were fortified to protect the eastern frontier. Diyarbakır was the heart of these operations on the Upper Tigris, its control stretching as far as Mosul, which in turn dominated the river until Baghdad. Husain writes that, with the Ottoman bolstering of the major nodes on the Tigris and Euphrates, the rivers became 'two of the greatest thoroughfares in Eurasia'.

Keleks, as remembered by Zayyat, were a cog in the engine of an imperial superpower as much as they were a family business. Istanbul placed orders via *kelekçiyan* – local specialised contractors – and provided these workers with raw materials.* Zayyat's ancestors were the *kelekçiyan*. Theirs was a time when the river was a whole, with rafts going all the way to Basra. Social and political movements matched the geography, and just as each part of the Tigris and the basin that it drained was connected to every other, so too were stations and fortress-cities bound together.

But the river no longer looked like a highway of industry. Zayyat's business was now fishing, limited to this patch

---

* In 1734, 'the imperial administration supplied the *kelekçiyan* of Diyarbakır with as many as fifty thousand skins for the construction of three hundred rafts'.

between two weirs. 'It's a shitty existence,' he said. 'When the railways came, it started the death of the *keleks*. The river is getting lower each year now, and no one cares. It's better to think about the past.'

It was in Fishkaya that our team had its first heated discussion. I maintained we should try to spend more time on the river, and Bişar threw up his hands, though the smile never fully abandoned him. 'And how will you do that?' he asked. 'The river is blocked every quarter of a mile.'

He was right. To the north of the city, near a village called Hantepe, we had seen the huge gravel-mining operation on the riverbed. Construction machinery drove around on islands of their own making, digging out ground around them so the Tigris became sinuous and, eventually, an unrecognisable mess of trickling channels. The aggregate that was dug out was highly valuable as the primary component of both concrete and asphalt. Its mining changes the shape of a river, makes banks unstable and affects biodiversity in the water and on nearby land. By some reckoning, sand and gravel mining is the largest extractive industry in the world. It is a plague on the Tigris.

As the river bent eastwards, we'd see the same again, and worse. There was no way a boat would make it. And then there was the security. Bişar and Angel were adamant that attempting to be on the river south of Diyarbakır was dangerous. 'We have permissions,' I kept saying, lamely. But it didn't matter. The government didn't mess around with foreign journalists in this region, especially so on matters of water. A couple of years ago, a photographer from a major international magazine was arrested and detained for his work on the subject. He'd been reporting on the very place we were heading to: the tentacled lakes that stretched out behind the latest, largest of the Tigris mega-dams, Ilısu.

We drove out of Diyarbakır, following the shape of the valley. A great mass of volcanic rock, formed from the lava of Mount Karacadağ to the west, forced the river into a new lateral movement. Our road was the highway that hugged the Tigris, surrounded by the rich floodplain of the plateau. Historically, the rainfall and snowmelt upstream would have caused the Tigris to burst its banks here. The soil would be inundated with alluvium that came from the limestone mountains which gradually, over many thousands of years, was ground down until the minerals could be carried and sprinkled like dust. The waning Tigris no longer escapes the banks so easily.

Around us fields of wheat, corn and cotton laid plush like a blanket. Bişar shook his head in the back seat and wondered aloud why thirsty crops were still grown so freely.

'They take water even when we don't have much. And then the farmers cover everything in chemicals. In summer, you can smell it on this road.' The soil, one satiated by mountain silt, was now drowned in invisible pesticides.

Tamarisk and willow grew by the river itself, making pockets of shade that must be a paradise in the summer. In rolling undulations, shepherds drove herds of long-eared goats. Paralleling the blacktop were rows of telegraph poles. At the top of each, all the way to the town of Bismil, were ragged nests of storks. Inside, safe from harm and suspended like medieval stylites, the eggs could hatch, and in September the families would fly back to Africa. Angel remembered once when one pair decided to stay and became so famous that people would drive out from the cities just to see them.

The first citadel mound we saw was at Üçtepe, sixty feet high and rising gradually off the plain. A little further on, another. These mounds were formed from layers of human debris collecting around a once-settled area. Particularly in areas where inhabitants built with mud-brick, as the Assyrians did, homes and buildings would regularly be partly demolished and rebuilt, because mud-brick deteriorates relatively rapidly. Each time this happened, the level of the settlement rose a little. Over time – centuries and millennia – humped hills were created. There are tens of thousands of them across the region, and in Turkey they are usually called *tepe*. At another, Ziyarat Tepe, the excavations have been backfilled, so little was visible to say that five millennia ago this was first inhabited, and two thousand years after that it became a provincial capital of the Assyrian Empire.

The river must once have coursed and crashed through this area at speed, benefiting from the gradient of the channel it had carved. One interpretation of the name Tigris is that it is derived from the Old Persian for pointed, *tigra*, and *tigri*, an arrow. The play of pointed arrow referenced the river's speed, which now seemed almost as forgotten as the language that christened it. Another idea is that the ancient Greeks took Tigris from the Elamites of south-western Iran, who used Tigra, which originally began as so much else with the Sumerians, and *Idigna*, a close approximation of 'running water'. For the time being, that at least was still accurate.

Curtains of mountains in the east rose to meet us, and we followed a peninsula of rock overlooking the first major confluence of the Tigris. Here the Batman River joined, concluding its seventy-mile journey from the Anti-Taurus mountains. It was a weak meeting. The Batman was dammed upstream and came in bifurcated from the north, separating around a large island of sand at the convergence. The Tigris also sneaked around a sandbank, hugging clear white rock below us.

A ramshackle village with more free-range goats than homes watched over it. The wind was fierce. At the water's edge, a beautifully constructed *kelek* bobbed up and down, pulling at its tethering like a restless pony. It had been made for communal use by a man called Mesut, who joined us now, wrapped tightly in an army surplus jacket.

Four tractor tubes gave the *kelek* buoyancy, and Mesut had carved railings on three sides and secured chiselled oars to the side. He'd even installed a small seat for the rower. Mostly it was used to cross to the main road on the north side, and then a car would carry villagers on to Batman for work or shopping.

The area on which we now stood had been flooded the year before, when the downstream Ilısu Dam was filled. But the water receded over the winter, and Mesut didn't know why. Once we had admired the raft for long enough, he led us back uphill to his home, stepping over hundreds of fish carcasses, no bigger than the palm of my hand. The water level had dropped so quickly that they were stranded and died here. Mesut then pointed to the crumbling soil embankment alongside the path. It was riddled with holes, and in each was the discarded, translucent skin of a snake. When the water first rose, thousands of black snakes escaped uphill to the village. They came through the doors and walls and windows and plumbing. The villagers tried not to kill them, but it was all a bit much, said Mesut.

His son brought us tea, and Emily and Bişar played with baby goats. A calf licked Claudio's camera. Mesut showed us pictures of the village from the previous decade, and I thought of how often I'd been in this scenario, looking at a memory of somewhere now destroyed. Maybe that's always the way, that there was always a better time, and we're wired to look backwards. But this felt more than that, as if these disparate communities on the river were joined together by collective loss, and an uncertainty about how to move forwards. We finished the tea and left the village. Mesut sat on the wall of the cattle shed and waved.

Chapter Six

The Water Brought Them, and Sent Them Away

Day 8
Hasankeyf
River miles: 168

New Hasankeyf sits on a windblown patch of scraped earth, and from a distance the rows of identical houses look like a model village. Its position is artificial. Just a few years ago this was bare hillside, two hundred feet above the valley, devoid of anything interesting except perhaps proximity to the confetti of birds. But since the waters of the reservoir came, the level on which the population must live had risen, too. The power of the mountains had gone, the horizon narrowed and dulled, and the heavens and earth pulled closer together.

I never saw old Hasankeyf, and to Arif this was a tragedy. I met him in the new bazaar: a purpose-built oblong market in which all but two of the prefab units were still empty. He was in the only coffee shop, crumpled at a table. A face that once must have had great character seemed to have fallen as low as his spirits. He was rake-thin, and unwell, cold. I was cold, too. Even inside, there was no escape.

'It's because we're halfway up a mountain,' he said. 'People aren't meant to be here. It's always cold.'

Arif had lived in Hasankeyf for all of his forty-five years. His two sons were born there, and they worked in his textiles store where he taught them how to use a loom, just as his father had taught him. For much of the last decade, the threat of displacement had hung over them, and he still remembered the exact day when their lives changed for ever.

'The tenth of November 2019,' he told me. 'That's when the first water came to my store, and we moved up the hill. That's when Hasankeyf died.'

Hasankeyf was among the most famous villages in Turkey, even before it was drowned. It was a capital of the Ayyubid dynasty, established by Saladin, the most famous Kurdish leader of all, and was said to be one of the oldest continuously inhabited settlements in the world. Those twelve thousand years of human history were seen in thousands of cave houses carved into limestone, and in hundreds of monuments from architects of the Assyrians, Romans, Byzantines, Ottomans and other empires who had claimed this labyrinthine complex. It was the only place in Anatolia that still boasted architecture from the Timurid dynasty, who blended their Turco-Mongol origins with a strong influence from Persia; later, they would form the Mughal Empire on the Indian subcontinent. They were represented here by a vast, fifty-feet-tall, fifteenth-century tomb with intricate turquoise glazed tiles.

Hasankeyf was a stop-off on the Silk Road. Until the flooding, a ruined twelfth-century four-arch bridge that once spanned the river, and probably carried Marco Polo and Evliya Çelebi across, was survived by two piers and an arch. At the heart of its beauty and success ran the Tigris. But that was the ancient past. Recent history saw a thriving commercial town, making the most of rich and well-preserved heritage, and welcoming visitors from all over the world.

'It was an outdoor museum, and it seemed too important to destroy,' said Arif. 'At least, at the beginning.'

Hasankeyf's downfall was the hydroelectric dam project at Ilısu, thirty-five miles downstream. Ilısu was a key part of the GAP, bringing electricity, irrigation and opportunity to south-east Anatolia. But it was to prove more damaging than any other component part of the GAP to date. Construction began in 2006 and, almost continuously from that time, faced opposition. Residents, environmentalists and engineers, local and international, came together to protest. They took the project to the European Court of Human Rights. Nothing worked.

An estimated eighty thousand people were eventually displaced by reservoirs stretching back ninety miles from the dam wall on the Tigris, and another 150 miles on its tributaries. Habitats and biodiversity areas that had become even more valuable after the damage to the Euphrates were now lost. Over three hundred archaeological sites were flooded in the area, and two hundred villages. Hasankeyf was the best known of these.

As I sat with Arif, I realised we were both wondering the same thing. How could it happen? He said he had thought about it every single day. We walked out together through the empty streets of new Hasankeyf. On the edge of the town, a handful of monuments had been relocated to the newly christened Hasankeyf Cultural Park. Most dramatic was the 1,100-tonne mausoleum of the Timurids. It looked lonely, disconnected from time and space, sitting squat on this dusty patch beside a piece of a Roman gate and a thirteenth-century bathhouse.

The docks, too, were odd, perhaps because they were new.

Beside a steel jetty were two boats decorated like pirate ships with crow's nests and Jolly Rogers. Like a dystopian Disney World, these were vessels for tourist trips which would presumably happen in the future. For now, there was no one. We got in another, simpler boat, with a tarpaulin covering wooden benches. A friend of Arif's came to join us, wrapped tightly in a scarf and quilted jacket, his face hiding under a tweed flat cap and behind a grey beard. His name was Ridvan, and he too had been born here.

For a while the skipper worked the wheel, manoeuvring this way and that to pull clear of the jetty. But each time the bow came out, the stern drifted back in. After a few moments I realised: he doesn't know how to do this. There was never a lake here before, so no one had experience sailing big boats. For centuries the vessel of choice was the *kelek*, as Evliya Çelebi and others noted, which crossed the river from side to side and travelled downstream to Mosul. Now the swift river was hidden until an inert ocean. The captain's desperate, hopeless wrestling of the wheel was a microcosm of new Hasankeyf.

Finally out, Ridvan sat beside me, his voice low and booming. 'There's a building there,' he said, pointing down into the water. 'That may have been the very first university in all the world.' A tower from it now sat in the cultural park on the bluff. The rest was beneath us, somewhere, but as I peered into the lake all I saw was the bruised reflection of a clouded sky.

Ridvan had a mental map of the layout of Hasankeyf preserved in his mind, and he signalled to various sites as someone might do from an aeroplane window. 'Beneath us now is the bridge,' he announced. 'The arch would be directly below us.' Arif joined in, indicating where his shop was. All of us stared pathetically over the side of the boat, flecks of water lightly spraying our faces.

Angel sat on my other side. 'I've been here a hundred times at least before the water came. And I have no idea where we are. It's like a new place.'

'Well,' sighed Ridvan, 'we are now passing history. And it's gone, and we'll all pay the price. If you don't know your past, how can you know the future?'

We floated over the bazaar towards a steep rock harbour. A behemoth of concrete encircled the mountainside behind. At the top was an artificial platform still covered with construction material. An excavator swung side to side behind a fence. Beyond, where the pale grey of the new ended, the beige and green of the old finally began. Hills rolled to the horizon, folding in on themselves as valleys and gorges. Switchbacked steps were cut into the limestone, entranceways fashioned out of overhangs, and everywhere were caves. They came in all sizes and riddled the land like holes through cheese. Grass grew in abrupt tufts between them.

There was life, still: a single shepherd on the horizon, and crakes, snipes and finches settling or bustling. But whatever was left was there against the odds, in a diminished homeland. If the Bonelli's eagles that once called this home were still here, they'd be somewhere in the distant high cliffs. But no one I spoke to has seen them in a while.

This area, once Hasankeyf's crown, was now all that was left.

'This whole valley was shops and markets,' said Ridvan, once more conjuring the past. 'Some of the shops would be in caves, a hundred and fifty feet or so below us now. So you come off the river, walk up here, with the rocks and caves stretching above, and then you could walk up the steps to the higher areas.'

'It was exhausting!' said Angel, remembering her visits as a child. 'We'd always complain if we had to climb.'

'There was a lot of industry here,' Ridvan continued. 'Oil, butter, candles, sesame. And everyone had their own special-ity.' Kurds lived on this side, and Arabs, Armenians and Jews further back. Ridvan paused and pursed his lips. 'And they didn't just flood this. They filled it in with concrete first.'

This time it was Emily who asked why. She had barely lifted her camera, and I knew she was shocked by this place.

'Because this is Kurdish history,' Ridvan answered. 'Water can cover it, but cement can destroy it. This way, they slaugh-ter the history of the Kurds. But it's actually everyone's history.'

This had been a major source of contention from the local population. The town, like the region, was predominately Kurdish, and much of the heritage connected to the Kurds. The Turkish government has denied all accusations that the project is anti-Kurdish and maintains that it will strengthen the economy in the region for all. But the reality feels different.

The reservoir has displaced Kurdish communities, flooded sites significant to Kurds and blocked off historic routes used by Kurdish nomads. The dam has a life expectancy of fifty to seventy years. As with everything related to Ilısu, the concrete poured on the old town was deemed excessive, and made sure that nothing could be recovered even if someday the water drains out. Ridvan used the term 'security dam', as Bişar and others had repeatedly since we began our journey on the Tigris. Ilısu undoubtedly brings the Turkish govern-ment greater control of the mountains once used by the PKK and offers leverage on countries downstream. That, alongside the maintained aims of power generation, has given Erdoğan enough support domestically and politically to persevere without compromising in the slightest.

A path led us up, under a Roman archway, into a warren of caves. All were shaped carefully to protect from the wind, with small entranceways and large interiors. 'We called these bureaucrat caves,' Ridvan laughed. 'The rich people lived here. As a child we'd come and ask for candies at Eid.' He smiled now, for the first time since we met, and skipped over the rock in a way that belied his six decades. But his face dropped again, and he was brought back too soon. 'Now all that childhood is under the water. The Dicle was so fresh and beautiful before. Humans came here for the water. And now it's because of water that they have to leave again.'

He led us past the skeleton of a church, where ornate crosses were carved into sandstone. A bell tower still stood proud, shorn of its ability to call even an absent faithful to prayer. The building had been used by the Assyrian Church of the East. The ethnic Assyrians, Chaldeans and Syriacs who once worshipped here, and who we would meet as we moved south, considered themselves direct descendants of Assyria. To them, this area of their ancient civilisation was a homeland. In Syriac it is called *Beth Nahrain*, meaning 'between two rivers', and roughly corresponds to the geographical understanding of Mesopotamia.

'This is it,' shouted Ridvan, ducking out of view beneath a crumpled hillock. We followed and found him crouching in the doorway of a cave. 'This was where I was born.'

I bent forward from the tailbone to avoid scraping the roof. Fine rubble covered the floor, but the walls and domed ceiling were solid. 'I slept in here,' said Ridvan, pointing to a small alcove. 'We were four kids, and I was the youngest, so I got the smallest bit. The children were beside the fire.'

There were six people in the cave including his parents, and at the back a concrete partition created an illusion of privacy. The walls were dotted with niches for clothes and food. 'These small ones here were for candles,' Ridvan said. From the roof a hook had been carved to hang meat and yoghurt. Ridvan's thick fingers stroked the smooth, black rock.

'I loved the mornings then. I'd always wake to see my mother shuffling to the fire, and then I'd listen to the kettle boiling.' He perched in a recess. 'I can't describe the feeling of being back here. It reminds me of a happier time.'

Ridvan had moved to Batman a few years previously. It broke his heart to leave, but it was worse to stay. He had led campaigns to protect Hasankeyf and petitioned local government. He ran for office in Batman and gathered activists to protest when the construction machinery arrived. In 2012, he was arrested and jailed for a year and a half. They made an example of him, he said. Another friend of his, who had first told me about Ridvan, now lived in exile in Europe. So many lives had been changed by the dam. Ridvan was forbidden from leaving Turkey, but when his travel ban ended he wanted to see the world, just as the world had once come to see Hasankeyf.

We took the boat back to new Hasankeyf and ate a kebab in the only functioning restaurant in the bazaar. A portrait of Atatürk hung on the wall, and when I asked about it the owner shrugged and said, 'He wasn't nearly as bad as what came later.' Ridvan drove home and Arif shuffled back to his house across town. We checked into the only hotel in town. A group of students lounged around by the entrance. They were here to see the cultural park, they said. None seemed interested. On the stairwell were cheap prints of old Hasankeyf, hanging in a building that existed only because the heritage they depicted had been flooded. The irony seemed lost.

Before Ridvan left, he'd told me that this area was ideal for wind and solar power installations. Now, as hydroelectric dams are being re-evaluated worldwide and coming under scrutiny, it would seem wise to invest in alternative renewables. But that wasn't the point. The GAP project had survived even as all but one of the international firms involved pulled out. Even as the budget spiralled to two billion dollars. Even as opposition, diplomatically and from public figures like Orhan Pamuk, increased. The dam was built, the valley flooded. Heritage was lost and residents moved. The Euphrates soft-shelled turtle, already absent from the Euphrates itself, was now likely to become extinct in Turkey. The town had met nine of the ten criteria for inclusion on the UNESCO World Heritage List. Only one is needed, but a submission can only be made by the government, and Turkey refused to.

Chapter Seven

Runaway Monk

Days 9–11
Ilısu | Tur Abdin | Sulak | Cizre
River miles: 233

We drove south-east on mountain roads that wound around hillsides like thread. At a high point close to Ilısu, we scrambled over an embankment to get a view. It was near enough to see the huge DSI emblazoned on the rock wall and make out the shapes of individual boulders enshrined in concrete. The way we watched from a distance, and how the landscape seemed cowed by the structure, water waiting patiently at its gateway, gave the site a strange sense of reverence, as if we might be reconnoitring some ancient tomb or castle.

Beyond it the Tigris cut deep, narrow canyons, engraving itself on the land. It is famously swift and capricious here. During E. B. Soane's journey from Diyarbakır to Mosul by *kelek*, he wrote of this section:

The river pursues a remarkable course here. The reaches are straight and short, and owing to the similarity in colouring of the opposite banks it is impossible to see the turn – often less than a right angle – till right upon it.

Huge hills rise up beyond their lower slopes covered with trees, and above all we could see snow-capped peaks. In these wild gorges, of a beauty of spring verdure, of a magnificence indescribable, we felt – as in all effect we were – but a chip swept along the great river.

Much of this drama is because of the throng of sprawling limestone mountains to the south that form a vast plateau, reaching out almost to the modern border with Syria and then dropping, dramatically, in a series of sheer, sharp cliffs. This is the range of the Tur Abdin, historically considered as the northern border of Mesopotamia.

To the armies of antiquity, the Tur Abdin would have been formidable. The ancient Assyrians approached from the south, Romans and Alexander from the west, and all found it to be a natural fortress for their enemies; a treacherous place to fight. The name means 'mountain of the servants of God'. For centuries it has been an important Christian enclave and formed the heartland of the Syriac Orthodox Church. The Syriacs themselves may well have been one of the first peoples to adopt Christianity.* Protected by the fierce rock formations, with escarpments as outlooks and natural moats for protection, there were once more than eighty villages and seventy monasteries hidden here. Thick oak forests offered still further camouflage, and each settlement and spiritual centre was built in such a way that they seemed to become part of the plateau itself.

During the First World War, between two and three

---

* Designations are complex because of the many differentiations and divisions which have developed in the seventeen centuries since the Syriacs have been in the Tur Abdin, but here the term Syriac can be used clearly to refer to the peoples who mostly belong to the Syriac Orthodox Church, who are considered as a distinct ethnic group, speak an Aramaic dialect called Turoyo and perform liturgy in a dialect of classical Aramaic, also called classical Syriac.

hundred thousand Syriac, Chaldean and Assyrian Christians
were murdered in a massacre known as the *sayfo*, or the
'sword'. Much of the slaughter occurred during a six-month
period in 1915, concurrent with the Armenian genocide in
which as many as a million to a million and a half Christians
were killed. Together they form the most shameful chapter
of the Ottoman Empire, which the modern state of Turkey
still denies. The Armenian attacks were systematic and wiped
from the landscape almost any trace that they once peopled
this region. Most of the Ottoman death squads, and the local
Kurds who carried out many of the attacks on their behalf,
made no distinction between the Christian denominations,
and the Syriacs were victims of this broad sweep of ethnic
and religious cleansing.

Things have not improved dramatically for the population.
Under Turkification, Syriac land was reappropriated and
renamed in Turkish. Oppressed Kurdish tribes, driven from
their own lands to the east, arrived in the Tur Abdin where
they clashed with Syriacs. Then the PKK was established, and
Syriac communities were stuck in the middle. Thousands left
for Europe, and now there are only fifteen hundred to five
thousand Syriacs in Tur Abdin, centred around a handful of
embattled monasteries.

Mor Gabriel, founded at the end of the fourth century, is
the spiritual heart of the Tur Abdin and the seat of the bishop.
There are thirteen resident monks and three nuns, making it
the largest surviving Syriac Orthodox monastery in Turkey.
We detoured away from the river by road, but were headed
to another site, even older. On the southern rocky wall of
Mount Izlo, looking out across the plains of Syria, is the
monastery of Mor Augin, built into and onto the mountains
so the two become one.

It was founded in the middle of the fourth century,

perhaps as early as 361, by the itinerant Egyptian St Augin, considered by some as the founder of Syriac monasticism. A greater number would scorn this suggestion because the accounts of his life came at least four hundred years after his death. The story goes that after an early life spent pearl diving in the Red Sea, sharing his finds with the poor, Augin joined the monastery of St Pachomius in Upper Egypt. Initially he was a baker, who worked occasional miracles on the side, but soon he began to gather a brotherhood.

He left for Mesopotamia with seventy of these monks in tow and settled on Mount Izlo with the view over the plain. This was the very eastern extremity of the Roman Empire, which was in the process of adopting Christianity, and thus the monastery attracted their support. There is a possibly apocryphal statement from the Emperor Constantine that names Augin as one of three great spiritual warriors.[*] Monasteries inspired by him grew across the east, from Egypt to Persia, and after his death in 363 he was buried in a cave under the southern altar of the monastery. For those Syriacs who believe his story, which is certainly the minority, he is called the 'Second Christ' and the monastery the 'Second Jerusalem'.

Intrigued and a little confused, we followed a track leading up to an imposing wall of rock. Just visible, suspended by faith alone perhaps, was the beige brick of the monastery. It was two or more miles away and inexplicably there was a locked gate across the road. The road itself felt incongruous; the rest of the approach was bare, rocky terrain. Beside the padlock, a sign read 'Mor Evgin' and gave a list of opening hours. Clearly, something had changed.

---

[*] 'These three warriors are known in our kingdom: Antonius in Egypt, Illarion at the coast and Mor Augin who moved out of Egypt and come and settled down in your region and enlightened it. We plead and beg of him that he prays for us in front of Our Lord, so that we and our kingdom will be protected and safe.'

We walked, each taking a different path through the boulder garden. A balcony in the distance lay supine on the rock and a bell tower stood sentinel. We climbed the two hundred winding steps of a stone staircase and, at a gate that looked more like a portcullis, Angel yelled again in a very non-spiritual way until a middle-aged man arrived. They argued for a while, and eventually she guilt-tripped him into letting us through on the understanding that we'd just take a brief rest, and no photographs. 'We just need water,' she pleaded. 'Who would deny thirsty travellers a drop of water?'

Behind the main complex, a series of orchards stepped up the cliff. There were suggestions of monastic cells higher on the mountain, now disused. Three hundred and fifty consecutive priests had watched over this place, with thousands more studying and praying under them, until a pause in 1970. Seventeen centuries of unbroken devotion had left their imprint on it and still vibrated through the monastery as we walked, the rock under our feet taking on the rhythm of a chant.

Inside the church, the oldest part and spiritual heart of this holy place, we stood together, dwarfed under a domed roof. The plaster on the nave had been applied by some of the earliest adopters of Christianity and still it stuck fast, only peeling slightly at the corners. The austerity was beautiful. Three vaulted windows high above lit the chamber, and at the altar a modest lectern buckled under the weight of a bible. I felt it among the most sacred spaces I had ever been to. Emily and I drew close, arm in arm, grateful to have been brought to such a place.

There was access to a cavern below the church via a winding passageway, and I lowered myself down. Here was the tomb of Augin, somewhere under the rock, and spidery

Syriac script crept around the walls in webs of prayer. In an alcove, two icons of the Virgin Mary rested in the bosom of the limestone of the mountain, dried wax surrounding them like a moat. I lit a candle and stood in its glow, head bowed first from the shape of the space, then from everything else.

I found some of the others by the balcony, where Angel was stopping Bişar from trying to ring the bell. He had a silly grin on his face and was enjoying himself. It was the largest bell in Turkey, he said, and I wondered how he'd come to this information. Suddenly, from a tiny doorway that none of us had noticed, a monk burst out. He was young, with dark eyes and a long black beard. His tunic floated loose over a slight frame, and his eyes darted in fear when he saw us. Briefly he considered returning, but the door had closed behind him. I introduced myself.

'I have no time,' he said. 'There are things to do.'

Angel asked gently if we could talk, and the man looked terrified. Give him a few minutes, he said, and he'd be back.

He disappeared into the maze of passages. I sat on the wall with Angel and Bişar and looked out across Mesopotamia. A few moments later Emily appeared, too. She had also run into the monk and been dispatched equally rapidly. 'He has the manners of someone who's come here to avoid people,' she said. After a few moments, the sound of swinging iron carried up on the wind, and we looked down to see the monk leave by the main gate. Steadily, he descended stone steps below us. At the bottom he continued, now on the tarmac. For ten minutes we watched until no doubt was left. He was escaping, very, very slowly. After the centuries of persecution at this place, I felt bad that it was our visit that had finally sent the monk fleeing.

With little else to do, and feeling that to give chase would

be both inappropriate and of little use, we watched his pur-
poseful strides to the locked gate on the access road. I was
intrigued – then what? But he had a plan, of course. A movement of arms and he produced a phone. Within ten minutes,
a car had arrived. He climbed inside and was gone. We'd
lost the monk.

We walked back out, bemused, and looked in a few other
empty rooms now that the priest was gone. Behind one
door, more modern than the rest, nine Germans were having
lunch. A young man stood up abruptly. 'Ah,' he said. 'You
can't be here.' We were shepherded out and regarded each
other in the light. The man was in his early twenties, handsome and broad. He had come from Germany to study at the
monastery. He was learning classical Aramaic. His name was
Marcellus. I asked about the monk.

'That's Father David,' he said. 'He's from Greece, and a
great and holy man. I learn a lot from him.'

I said I'd just watched him run away.

'No, no, he's gone to pray in the church.'

I told him about the walk to the gate and the phone
and the car.

'He's praying inside,' said Marcellus, still smiling serenely.

Marcellus spoke for a while about the importance of the
place, and of how the monks watched Syriacs being massacred
in the valley in 1915. It was better not to have visitors. Some
places should only be for those who can understand them, he
said. I disagreed, agreeably, but he continued, always smiling. His family was visiting, which was okay, because he was
a student. 'But you should leave,' he said. 'Normally when
there's a locked gate, it means someone doesn't want you to
come inside.' He beamed.

We left the second Jerusalem, more confused than when
we arrived, and followed the path taken by Father David like

reluctant pilgrims. At the gate, we looked one last time at the tourist sign with the visiting hours and drove away, leaving the Tur Abdin and the monks on the hill.

Before we rushed out of the mountains, like the river itself, we walked to a high point near a village called Sulak, where a winding shepherd's path led past an orchard of fruit trees. A young man, another friend of Bişar's, led us wordlessly until the path fell away. On the lip of a great canyon, the cobalt blue of the Tigris coursed through far below.

Our side of the gorge had oak and willow and promontories of rock, and the far side exposed to the wind, unshielded, unwooded. A bird of prey, perhaps another kestrel, glided across as if on a string. For the first time in days we could look back across the mass of uneven mountains and see only bluffs and hardy trees. I had become used to every strategic point being occupied by concrete squares and military antennae of army barracks.

The river curved dramatically from the north and then, like a sailor steadying a ship in high winds, its course gently corrected. Underneath, a mass of black moved as one up a gentle incline, softly bending and sweeping. In time it took form as a herd of sheep, heading towards dark hair tents of the nomads, like a murmuration on land. These hills had been home to the PKK for decades. Sulak had been at the centre of Turkish military operations to root them out even recently. And yet it was hard to imagine any hardship or danger in somewhere this serene.

We poured tea from a flask and watched as the falling sun cast the canyon in ever richer light. Bişar's friend sang us a song about the folk hero Alo Dino, who took from the rich and gave to the poor. He tapped a wooden staff on the rock, and the sound met the timbre of his voice. Both flew out across the valley, soaring about the river, and drifting out of mountains with the squeezed Tigris.

Alo Dino was a Kurdish Robin Hood of the sixteenth century, and he stopped *keleks* in these gorges. If they carried anything of great value, he'd take it and distribute it to the needy in close-by villages. As his fame became well known a local *Mir*, or tribal head, became determined to catch him. The *Mir* put a beautiful girl on a *kelek* as bait, and when Alo Dino stopped the vessel they got drunk together. The *Mir*'s men were waiting. Alo Dino was imprisoned and tortured, and the *Mir* would visit and taunt him. 'Is there anything worse than this?' he asked Alo Dino finally. Chained to the wall, scared and bruised, Alo Dino replied: 'What's worse is if you don't give milk and tea to strangers if they come to your home.' It was a very Kurdish story, where a lack of hospitality was deemed worse than torture. The *Mir* cried, suddenly realising that in front of him was a good man, but Alo Dino died and there was no redemption.

As the Tigris spills from the mountains, it helps create Al-Jazira, or 'the island'. This was the name the early Arab geographers gave to the land sandwiched between the Tigris to the east and Euphrates to the west. By the time the two rivers reach Syria at Al-Jazira, well over two hundred miles separate the watercourses. This is probably the basis for the mythology in both Turkey and Syria that the Tigris is a male, phallic and straight, and the Euphrates is feminine, boasting the curves of a woman. It was not, as one of the *dengbêj* told me, because a man's opinion is direct and to the point and women can't make up their mind.

The modern city of Cizre* meets the onrushing Tigris close to both the border with Syria and Iraq. The same medieval Islamic scholars who named the region believed that Cizre was founded by Noah after the great flood, and that the ark came to rest on nearby Mount Judi. From the roof of the tall and well-positioned Hilton hotel, where we had to promise that we'd buy a coffee to allow entry, we looked over into Syria through a mess of walls, wires and watchtowers. The river became stretched thin, separating like strands of string to create long, narrow islands of bog. On the far side of the border stood an old bridge, now connecting nowhere to nowhere, stranded in no man's land and distanced from the water. The hotel manager told us that a few years ago the Turkish had re-routed the water to bring the bridge under their authority, but now it was back in north-east Syria. The Turkish army sometimes shot at people who climbed on it, he mused. But why were we interested, he asked suddenly, remembering his manners. I wasn't quite sure how to tell him we were going there next, and ideally hoped to stand on top.

---

* Cizre was founded in the ninth century by the Emir of Mosul on a plot of land that the Tigris surrounded on both sides in spring, so it too was called the island, Jazira, with Cizre the Turkish form.

Chapter Eight

Land of Acronyms

Day 12
Borders | Ain Dewar
River miles: 236

Angel and Bişar returned to their families in the rental car. As Turkish citizens, it would have been almost impossible for them to enter Syria in the way we planned to do. Emily and Angel had become close, and I knew Emily would miss travelling with her. I hope this journey won't feel like 1,200 miles of goodbyes, she said.

It took two days for us to reach the old bridge. In a borderless land, the same journey would have taken just a few minutes. Instead a Turkish car took us to the border where we switched to one with Iraqi plates, and for a few hours we moved between booths handing over passports and chits and forms. Once through, we were driven back east to the crossing over the Tigris into Syria, where pontoon bridges moved buses of passengers back and forth between the regions.

The Kurdish name for the area is Rojava. Historically, this was a blanket term that covered all the Kurdish majority regions of Syria. Officially, it is North and East Syria (NES) and is administered by the Autonomous Administration of

North and East Syria (AANES.) The administration was declared in 2012 after the outbreak of the Syrian civil war led to the expulsion of regime troops from the area, and the beginnings of self-organisation and de facto autonomy.

We were now mired in a land of acronyms, which to me seemed a successful way of making each aspect of the place seem cold and detached. The political assembly of the AANES is the Syrian Democratic Council (SDC) and its military the Syrian Democratic Forces (SDF). The SDF are a coalition of various militias from different ethnic and religious backgrounds, led primarily by the Kurdish People's Protection Unit, or YPG. The YPG has a sister militia for its female fighters, the YPJ. In 2015, the YPG pushed ISIS back from Kobani, at the time under siege, and fought alongside and with the support of the US-led multi-national military, the Combined Joint Task Force – Operation Inherent Resolve (CJTF-OIR). Despite close ties on the battleground, and the SDF's key role in defeating ISIS, the YPG are seen by the Turkish government as inextricably linked with the PKK.

The Syrian regime maintains some presence in the region, and there are Russian and American bases and patrols. The crossing is possible from Iraq based on a mutual understanding between Kurdish-led administrators, and at the border we went through a familiar rigmarole of makeshift offices and concrete waiting rooms. Around us, NGO workers, foreign and local, made the same dash, and the lilt of Kurdish echoed around bare walls with Arabic, French, English and German for company.

Amid the chaos stood Sam, a six-foot-four, ginger-bearded, Arabic-speaking Catholic from Montana who somehow seemed right at home here. I had forged a friendship with Sam in Erbil at a British-style pub quiz on Monday nights, followed by karaoke in a Filipino-run Irish bar. He

had lived in Damascus a decade ago, studied in Beirut and
now ran his own NGO in north-east Syria, as well as writing
about the region's Christian communities. He had come as
our translator. Most importantly, he was an island of peace
in an ocean of stress.

We had been travelling for almost two weeks. Most nights
we had camped or slept on floors. We were up late, out
early and ate sporadically, in the feast or famine style that all
travellers will know. As far as long, uncertain journeys go, it
was what we might have expected. What had surprised me
was the anxiety we all felt moving through eastern Turkey.
Emily had taken far fewer pictures than normal. Claudio
hadn't been allowed to bring a drone into the country and
was irritated by being pulled away from the adventure on
the river itself. It was only when we reached Syria that we all
began to relax a little.

The administration in NES was generally welcoming of
foreigners and journalists, said Sam. They were keen for all
the press they could get. On the Syrian side, an official tucked
an entry form into our passports, and lamented the fact that
we'd miss spending Assyrian New Year there. Would we
come back the next year, she asked, for fun? The last stretch
of detour to reach the bridge took an hour and a half to
cover twenty miles on gnarled, unsealed roads. Alongside,
in crumpled hills that flanked the road, a warren of holes
burrowed all the way from the border into the hinterland.
This network of tunnels was part of the defences of the YPG.

In October 2019, three days after President Trump ordered
the withdrawal of US troops from NES, the Turkish military
launched airstrikes on the border, before moving in ground
troops who took control of sixty-eight settlements in nine
days. President Erdoğan's plan was to expel the SDF, create
a twenty-mile buffer zone on the Syrian side of the border

and use the annexed land to resettle some of the 3.6 million Syrian refugees in Turkey. Ever since, the population of the north-east, especially close to the border, had been on high alert.

The Zangid Bridge stands alone, the tip on the nail of a finger of Syrian land that pushes back against a Turkish border wall. The road ended abruptly, and we walked the last stretch over fresh, thick grass. Beyond, the drone of the Cizre highway blew across the international border. Just a single arch of the bridge remained, though at one time there would have been at least three. Black basalt bases of the others still stand like tree stumps. Each brick was broad, as long as my arm, and pressed in were limestone panels which shone in contrast to the basalt. Now a makeshift grid of iron girders protected them, but behind were signs of the zodiac, carefully carved and shaped in soft, grey rock.

There were eight panels, each three feet square, depicting astrological symbols in pairs. In one, an armoured horseman had once carried the severed head of Medusa in his left hand, though now the upper panel had been removed, probably looted, and only the horse's body remained. I had seen the original in a photograph taken by Gertrude Bell just before the outbreak of the First World War. Otherwise, the reliefs were in good condition.

In antiquity there was a Roman road that came from Antioch, crossed the Euphrates and arrowed towards Mardin, eventually ending in Cizre. Perhaps this is why some that we met on the Turkish side, and other accounts online, believed this bridge to be Roman. In fact, there is no proof that it was built at that time, and it seems more likely from the Arabic sources that it is a twelfth-century construction, from a time when this area was busy with traders moving laterally across

the fertile foothills. The river then would have been subject
to vicious flash flooding, and all but the most robust bridges
would have been washed away. Even until the end of the
Ottoman era the Tigris still flowed here with vigour, and
accounts of the bridge in previous centuries describe waves
and whitewater as the river rushed through.

Said Mahmoud guarded this and several other areas of
archaeological interest in NES. He was a colleague of a friend
of Sam's, and rode his motorbike between sites to make sure
nothing was looted or vandalised. He lived in the village of
Ein Dewar, on the promontory that overlooked the bridge.
A weathered face spoke of much time spent under the sun.

The sun moved around the bridge during the day, he said.
Perhaps this was to allow the images to be seen in sequence.
As he spoke, the limestone of Capricorn shone with reflected
sunlight, though it was dim behind the bars. It had been a
busy morning, Said told us. A Russian military patrol had
come through Ein Dewar and kids had thrown rocks at it.
Then the Americans came through and children cheered.
Neither of these things were unusual. It was well known,
he said, that east of the Euphrates had American influence,
and the rest had Russian, but both patrolled in the north-
east, and often encountered one another on minor roads.
The Americans were still popular, despite Donald Trump's
actions in 2019.

'It's not trust,' said Said when I asked why. 'But people like
them. Sometimes the villagers bring roses.' He knew they
were doing it out of self-interest. 'But their interests can save
us,' he said. 'It's much better than the Russians. No one trusts
them at all.' After four decades of control from Damascus, he
believed things couldn't be any worse. 'Even if the Americans
are just here for oil, they could take half of it, and leave us
half, and we'd still be better off.'

Said pulled himself up onto the bridge and reached down to haul us up behind him. I paused, thinking back to the hotel manager in Cizre. The border guards on the Turkish side would definitely be watching, confirmed Said. 'They'll know that you came from Turkey.' That was a little disconcerting, but he maintained we'd be fine. I let Claudio go first.

The surface was covered in small rocks to protect the dirt layer that covered the basalt. We scrambled to the apex of the arch and looked out. 'For a few years, we couldn't come here at all,' Said told us. Sometimes the Turks would open fire at his village, or at people having picnics nearby, and definitely at anyone near the bridge. Then something changed – he wasn't sure what – and only the north side of the bridge was off-limits. Then it changed again, and now the whole thing was safe. Relatively. He chuckled. I asked how long it had been accessible.

'Since two months ago. An American patrol came, and they climbed all over everything, and since then no one has been shot at.'

It seemed a loose way of measuring safety. If it had changed so quickly before, what was to say the sight of us casually strolling around the bridge, with cameras and notebooks, wasn't to be the next chapter in the story? Claudio smiled broadly when I mentioned this. He stayed on the top for half an hour with his camera pointed towards the border.

'It's been a hundred years since the river went under the bridge,' remarked Said, back on firm ground. He mentioned again the recent Turkish effort to reroute the water. The bridge, although no longer of any practical use, was still a useful symbol. But efforts to change the river's course had failed, and the bridge stayed in Syria.

A family spilled out of a people carrier behind us,

sprawling onto the grass with picnic rugs and shisha pipes. A teenager called Mohammed spoke to me in English. He learned from watching *Kill Bill*, he said. Other movies, too, but mostly *Kill Bill*. He liked the way they spoke. He told me this was a place for boys and girls to hang out and relax; few people came for the architecture or the history. He felt it was safe as long as he stayed back from the river. Peace and quiet is exactly what he and his friends came here for, and it made sense in an odd way. This was the furthest corner of their country that they could access. In our short exchange, Mohammed used the word motherfucker six times.

Another car had brought young, secret lovers. He had been in the YPG for seven years, and she was at university in the city of Al-Hasakah. They came here to be together. I thought of the irony of travelling to a place under constant surveillance from a hostile country to avoid prying eyes of family. Emily sat with them a while, drinking Argentinian maté and talking about love. She was good at this. She laughed more easily than I did, smiled more. I was a good listener, she would tell me. That was a useful quality, too, but it took more time to build trust. Almost everyone we met seemed immediately to feel at ease around Emily.

In the village of Ein Dewar we sat outside Said's house for tea. The walls and yard were made with the same scraped cement finish as the neighbours', but his home seemed spacious and had an aromatic garden. Said's eldest daughter and young son sat with us. Two nieces ran into the long grass with Emily and lay down among yellow cole flowers. Their hair was shining orange, skin pale and freckled. They giggled and wound flower stems between their fingers. Soon other children from the village joined, and Emily chased them through gently swaying grass. For a while they all rested in the shade of a chestnut tree, then began again.

Said's wife came out to see the cause of the noise. 'It's nice to see happiness,' she said. 'We go to bed with no electricity, wondering if Turkey is going to come back. It feels good to hear laughter.'

Chapter Nine

Fragile Existence

Day 13
Lower Mezri | Upper Mezri | Derik
River miles: 241

The Euphrates provides over 80 per cent of the fresh water in Syria, and unsurprisingly Syrians pay closer attention to it than the Tigris. In the last half decade its downstream water flow has reduced by almost half. In large part, this has been due to dam building at the headwaters, much like the Tigris. Challenges of pollution, water–intensive agriculture and uncoordinated development are shared, too. The Tigris, which only passes through the country for a little over twenty miles, was associated with just two things in Syria, Said informed us. It was, firstly, an uncrossable boundary. The second was for its aggregate. The gravel was given the name 'Dijla', after the river, and was sought after for construction across the region.

An unsealed road followed the Tigris from half a mile inland, and riparian villages nestled at the end of arterial tracks. Not long after the start of the civil war, in 2012, Turkey cut access to the river and adjoining fields on the Syrian side. Before that, it had been the primary source of irrigation, but for those first years of war many farmers could not buy seeds and fertiliser, so the lack of river access made no difference; they couldn't plant regardless. Now they had access to raw materials again, but the moratorium on river use forced them to irrigate from more distant, less plentiful streams, which had begun to dry up rapidly in recent years.

In a village called Lower Mezri we sat around a plastic table in front of the home of an elderly man called Abdul. He was wrapped in a thick coat and twirled the ends of a prodigious grey moustache as he spoke, and the two friends who had been drinking tea with him nodded politely. He liked to keep the building between him and Turkish soldiers, Abdul said. 'There's not a house here that doesn't have a hundred bullets in it. We don't turn the lights on at night and try to keep our heads down.'

I couldn't understand the logic. 'It's just because we're Kurds,' was Abdul's explanation. 'If the Kurds built a town on the moon, they'd want to attack that, too.'

Identity and belonging were at the heart of everything here. Abdul told us how his family came to Lower Mezri in 1929 after a Turkish massacre of their tribe in their original home, a hundred miles away to the north-east. He had heard there was a growing recognition of the Armenian genocide. He wanted acknowledgement of the crimes committed against his people, too.

Then he spoke of the Tabqa Dam, close to the city of Raqqa. When it was completed in the early 1970s, displaced Arab tribes were moved to the north-east and given Kurdish land. They were called 'the Submerged'. Abdul looked at the taller of his two friends and smiled. 'He is an Arab,' Abdul told us, and the man mumbled something about them all being friends now.

To visit these villages we had hired a tired minivan, with worn-out suspension and faux-leather seats. The driver, Rabi, was Syriac, and also came to us through friends of Sam's. Now, while we spoke with Abdul, Rabi stood under an Aleppo oak, arms folded, face tight.

'He noticed a stone from the altar of an old church here,' Sam explained. It was being used in the garden as part of a patio. These villages were all once Syriac, Rabi told Sam, until the populations were driven out during the *sayfo*. Syriacs like Rabi knew the history and occasionally spotted sacred places now reimagined or occupied by an uncaring or unknowing resident.

In Lower Mezri there were no Syriacs left. Rabi had a quiet word with Abdul, and a few minutes later came back to the minivan. He had taken a handful of soil, because this area was holy. Abdul came to shake all our hands, lingering longest with Rabi. They spoke again and seemed to have reached an accord. Abdul's Arab neighbour waved from the table. Suffering compounded suffering in this group; everyone a victim, at one time oppressed, massacred, outnumbered, displaced, sometimes by the ancestors of one another.

I had wondered with all the flying bullets whether people had actually been injured on this border. In the next village, Upper Mezri, a man called Mohammed showed us his foot. There were two black bruises between the ankle and the toes, and it was swollen and stiff. It had happened just over five months earlier. Mohammed was returning home by motorbike, following a farm track close to the river. He heard something, like a sound of air being sucked in, then a pop. He stopped the bike. A bullet had gone past his ear. Then dirt exploded by his feet where a second hit. There was no time to run. The third hit his foot, and all the flesh came out of the exit wound. He scrambled off the track and lay in a ditch for twenty-five minutes, bleeding out, until his younger brother carried him back to the village.

'Four toes were broken, and the foot is still useless,' he said.

His mother, Fatima, joined us. She wore a glittering abaya and pale blue headscarf and when Emily complimented her on it she smiled, but her face soon rested again in sadness. 'They lift their weapons when I go to the garden,' she said. 'It's no life here.'

Emily accompanied her to the threshold of safety, and Fatima held her arm, pointing out her herbs.

The family had lived on this plot since Fatima's father's time. She remembered, only a decade ago, when they could swim in the Tigris. Every house had a net for fishing, and when they went to the banks they'd see their neighbours and wave. There were Turkish guards on the far side then, too, but they just watched.

'Now they even open fire on children,' she told Emily. But the family would not leave; of that she was adamant.

'They'd have to kill us all,' she said, setting her jaw, and looking out across the smudged water to the dull grey walls beyond.

Three people had been shot in the next village, we heard, five in another. Three cows killed further on. We drove inland for the evening, not trusting the safety of a night in the villages. This was a good decision, though our visit was not without impact. The next morning Said told us that Fatima's home had been fired upon throughout the night.

'Because of us?' I asked.

'Definitely,' nodded Said. 'They counted sixty shots.'

I apologised and wondered what we'd done that caused such a response.

'It's not unusual for them,' Said added. 'They just asked that you tell people what they're living with. That's all.'

*

We slept in the city of Derik, which had grown in importance since the start of the autonomous administration. The outskirts were riddled with the YPG's labyrinthine defences, and any patch of earth that wasn't burrowed under was being built upon. Ribs of new constructions protruded from every side. It was still significantly smaller than Qamishli and Al-Hasakah, but they were two hours further from the border with Iraq. In Derik, many saw an opportunity. If north-east Syria continued to be governed independently, some reasoned, then the economy would grow, and NGOs and private-sector business would increase. The other argument was that, eventually, Damascus would re-establish diplomatic relations with neighbouring countries, and control of the north-east would return to the regime. Then there was the continued threat of another Turkish invasion. These opposing ideologies played out in tandem: entrepreneurs building residential blocks and shopping malls overlooking underground defensive positions.

The city centre was a tangle of concrete and iron, low-hanging electricity wires, scabrous road surfaces and beaten-up cars. In the bazaar, almost everything was new since 2012. Inside the contemporary incarnation of a Syriac Orthodox church were the remains of a fifth-century monastery. It was surely the oldest place of cultural significance in Derik, and well known as a site of miracles. The caretaker himself had seen the Virgin Mary many times, he said. There was also a dove that circled the church to protect it. We looked to the skies but saw only soft clouds and the sun behind them, pale as a coin underwater.

The owner of a fishing shop laughed out loud when I said we'd hoped to go down the Tigris by boat. They used to fish there, he said, but now they used inland streams. It

was just for fun, he said. The river and way of life associated with it were dead to them now. Fishing could only be
a hobby. Who was to blame? I asked. 'Take your pick,' the
owner replied, and returned to adjusting spinning reels and
arranging them on the walls.

Sam led us back through snarling traffic to the ornate
doorway of another Syriac Orthodox church. Kurds were
a majority in Derik, but there were still thousands of
Assyrians and Armenians living in the city's south. This
building was striking, with three large crosses adorning
a multi-domed red roof, and walls of sloping sandstone.
A carved archway swept over the stairs to the door, and
elderly congregants shuffled into a tall, open nave. There
were three rows of pews, crimson curtains with stitched
gold goblets hanging on either side of the altar. When the
doors were closed, the noise of the city's growing pains
was silenced.

Every day during Lent there was a service, said Sam. The
priest read verses in Syriac, while altar boys in bleached
gowns rushed around trying not to trip. One, slower than
the others, occasionally hoisted his linens to reveal grey
flip-flops branded 'Niek', with the trademark orange tick.
Emily sat among the women, who gave her a shawl to
cover her hair and held her hands during the service. When
Claudio tried to film them, they batted him away. I sat
with Sam, and no one dressed us or held our hands. None
of us could follow the Syriac, so we instead responded to
the cues of the congregation: standing, sitting, looking
holy. I thought of how much time I had spent in this region
trying to look pious. The altar boys lit candles, and swung
lanterns of incense, and then everyone stood up to leave.

At the doorway a heavy man in a deep-blue suit introduced himself as the deacon.

'Welcome to Syria,' he said in English, grabbing my hand in a sticky grasp. 'Welcome to the country with the greatest president and human anywhere in the world.'

For a moment, I was genuinely perplexed.

'Bashar al-Assad, our president and leader!'

He was a great warrior and humanitarian, the deacon said. He had defeated the terrorists, and everyone else, and soon would take back control of the north-east. He was a family man, and Syria lived deep in his heart. Did I understand all this?

I knew better than to ask what the deacon thought of the Autonomous Administration. A majority of Syriac, Chaldean and Armenian communities in the region supported the government in Damascus to a greater or lesser degree. They had been reasonably well looked after by the Ba'athists, and although there was a co-existence in Derik and beyond in north-east Syria, it was an uneasy one, haunted by what had come before. But even with that precedent, the deacon's views seemed remarkably enthusiastic. He gave me his card, pleasantly embossed in Syriac, Arabic and English.

'Come visit again some time. God bless you, and God bless Bashar al-Assad!'

Chapter Ten

The Waterkeeper

Days 14–16
Khanik | Faysh Khabur | Qasr Malataib | Khanke
River miles: 306

We waited in a packed holding area at the border, four miles south of the point where Turkey gives way to Iraq on the east bank of the Tigris. A crumpled iron roof provided some shade, but the weak sun of our early journey had now grown strong and beat down whenever we broke cover. It did not bode well for what lay ahead. By the time we had arrived back into the Kurdistan Region the day was half gone, and we shared a rushed goodbye with Sam.

Below Faysh Khabur, where the militarised border zone ended, Emily, Claudio and I walked to a bend in the river where Nabil Musa waited for us. Nabil was an activist and the best-known face of water protection in Kurdistan. He was part of a global movement called the Waterkeepers' Alliance, and for a decade had been the appointed advocate for protection of the Upper Tigris. Of roughly 350 leaders worldwide, Nabil was the first in the entire Middle East. He was in his mid-forties, strong and tanned, always coiled with a slightly manic energy. I had admired his work from afar since I first came to Kurdistan. He hugged us briefly, then got down to business.

'We should probably go,' he said. 'There's a lot of things here, a lot to do.'

With him were Salman Khairalla and Hana Ibrahim, who would accompany us all the way to the Gulf. Salman was born and raised in Baghdad, and the co-founder of an environmental organisation called *Humat Dijlah*, or Tigris network. He was a little younger than me, but had spent a decade travelling around Iraq and implementing projects on the rivers. Hana was half Kurdish, half Arab, with family from various places in the region, but had also grown up in Baghdad and come north in recent years. She was involved in civil society and environmental protection work, too, but her passion was animal rescue. She lived with a flock of disabled birds and regularly took emaciated animals off the streets of Sulaymaniyah city to find them a home. Both Salman and Hana were spirited in work and life. Salman wore bright turquoise glasses and had a wispy moustache, and Hana's shock of bouncing curls meant that although neither were much over five feet tall, they were both noticeable.

The banks were lush with thick grass, and wild yellow cole flowers shook in the faint wind, catching the light. In the water, two fishing boats bobbed up and down. Once they had been the same colour as the flowers, but now were jaundiced with age. A boatman called Molla Nawaf, found by Nabil in a nearby village, guided us in.

There were a handful of small villages, set back from the river or placed on overlooking crags. The first was Chaldean, the next Yazidi. Nabil shouted over the roar of our Yamaha engines that they had suffered heavily under ISIS. From a distance, on a cool afternoon in spring, there was a tranquillity here. I wondered how it had felt four years earlier.

For a few hours we saw no guns, no checkpoints. There was the noise of the engines but when they were cut, only wind and water. Taller hills, the Zinar, emerged to our right, and bottle-green fields of wheat, barley and lettuce opposite. The Tigris gave abundance to it all. Miles passed beneath the hull, and we felt each bend and bow of the river take shape.

On the eastern bank, in late afternoon light, we passed the remains of a home by the water. The concrete walls were still in place but a wooden trestle, once for grapevines, had collapsed, and there was no roof to be seen. Above, where the land climbed to reach fields beyond, an elderly man in ragged clothes sat cross-legged on a patch of grass.

The hermit didn't register our presence. Instead, he focused his attention on a plastic transistor radio cradled in his hand. It was not switched on. Nabil sat with him and asked his name. After a while the man looked up, eyes glazed, and mumbled a few sentences. Behind him was a Portakabin, and inside we saw shoes, torn clothing and bags of refuse: years' worth of overflowing detritus. Two dead mice lay at the threshold. Nabil remembered the man from the last time he was there, in 2013. He ran a water project in the nineties. 'From what I remember, his wife was killed by Al-Qaeda,' said Nabil. Then his son died. 'Now he's here, living in madness by the river.'

It was not a peaceful chaos into which he had slipped. The man seemed tormented, living in awful purgatory beside the water that he used to love. From his place on the grass, he switched on the radio but it played only static. He closed his eyes and listened. For days afterwards I thought of him, living there alone, year after year, the sounds of the radio and the river his only companions.

We camped by a water pumping station and the Peshmerga, the Kurdish armed forces, came to check on us. They wore red berets and arm patches that identified them as Rojava Peshmerga, from Syria, stationed here to protect the facility. The soldiers huddled round to hear where we'd been in their homeland. Later, a member of an intelligence department turned up, too, in plain clothes, and stayed longer than we would have liked. He ordered Molla Nawaf to take him out to fish. It was spawning season, and fishing forbidden, but this man had his own set of rules.

The lake began a full forty miles upstream from Mosul Dam. Where the river widened and sagged, our boats stayed close to Kurdish territory on the left bank. The right, to the west, was controlled by Federal Iraqi forces, and even though we had a valid visa and permission to travel, we could not cross here. Nabil asked Molla Nawaf what would happen if we tried. The lake was off limits at this time of year, he replied. 'So if they saw us, or anyone, they'd shoot.'

In the summer of 2019,* when the waters of the lake dropped lower than usual after a drought the previous winter, German and Iraqi Kurdish archaeologists found the ruins of a 3,400-year-old palace from the Mittani Empire. Cuneiform tablets were brought out and shades of wall paint found still in situ. We passed within a few hundred yards of the site, but Molla Nawaf was unwilling to risk a visit.

One village on the Kurdistan side had created a miniature resort by the water's edge, with a swimming pool and scruffy sofas laid out along the shoreline. This had originally been Molla Nawaf's village, Qasr Malataib, but his family's land beyond had been flooded by the reservoir. He, and Kurds of his generation, had lived through the Iran–Iraq War, then the genocidal Anfal campaign of the late 1980s when Saddam Hussein sent Ba'athist troops to devastate the north. Two thousand villages were destroyed during that time. I'd

---

* The same happened the year after our journey, in summer 2022.

seen the ruins of many in the region and visited more still that had been rebuilt. Between 50,000 and 150,000 Kurds were killed. Molla Nawaf remembered the attacks, the displacement, the chemical weapons. Sitting on a dusty sofa on the shingle beach, he recounted them as someone might talk about past jobs, or cities they'd visited. I wondered what it had taken to remove the pain from his speech, and where it went to instead.

Mosul Dam was finished during this period, in 1986, and seen by many as a vanity project to impress and distract the world rather than make any real difference to Iraqis. The Kurdish uprising followed Saddam's defeat in the Gulf War, and fishermen and pastoralists like Molla Nawaf simply moved with the times, readjusting to give themselves the best possible chance of survival. His life was hard, he said, and he was poor. He still came to the reservoir to fish, but there wasn't much left. The corrupt elite, like the man we'd met the night before, did what they liked, and everything worked on a system of *wasta*, or connections. Some idiots even fished with bombs, he said. They threw dynamite into the lake, then gathered the dead fish that floated to the top. The first time I saw him smile was when he thought of one man he knew who had his hand blown off when he tried this.

On our last night on the lake, we slept on an island. I asked Nabil about the lake, and how it felt to be here again. The last time he crossed it by boat was in 2013, when a flotilla of artists, environmentalists and activists had travelled in kayaks and traditional boats on sections of the Tigris in Hasankeyf, Mosul Dam and then south of Baghdad. It was the only attempt at a journey with similar ambitions to ours, and had gathered the loudest voices for river protection in Iraq. Some of them, like Rashad Salim, an expeditionary artist, and Azzam Alwash, the founder of an NGO called Nature Iraq, had passed on advice to us. Salman was with them then, too. Nabil had kayaked the Kurdish sections and thought back on what had changed in eight years.

'The situation is really bad,' he said. 'Worse now than then, and even worse every year. But what makes it even more so is not to do anything.'

He threaded the poles of his tent and looked across the reservoir. A setting sun lit the water like fire, spreading towards our island.

'We waterkeepers are the voice for this river,' he said. Nabil travelled around the region, educating those who lived by the waterways on how to keep their rivers healthy. 'Rivers bring everyone together,' he said. 'Every human being should have clear access to this natural resource. We should all have this right: air, soil and water. Without it we cannot continue.' The river he'd grown up on as a child had been destroyed by pollution, and he was determined to do everything he could to protect what was left in his region. It was surely a thankless task but Iraq would have been much worse off without him.

That night we ate bread and processed cheese from a sheet of plastic on the grass, and sat huddled close for warmth. In the distance, red lights flashed across the sky. Molla Nawaf said it might be tracer fire from the Iraqi forces fighting remnants of ISIS. He wished he'd brought a chair so he could watch it in comfort. He smiled again, for the second time. I thought once more about how little I sometimes understood of this place.

On our last day in Kurdistan, we brought the boats to shore opposite the island campsite. There, an IDP camp in a town called Khanke has become home to around fourteen thousand Yazidis. Most arrived after ISIS attacked Sinjar Mountain in Federal Iraq in 2014. The Yazidis are a religious minority in Iraq. Their faith has some shared elements with Abrahamic religions, but more still that is mysterious to outsiders. Their holiest site is the temple at Lalish, only thirty miles from Khanke, where three mountains converge to protect the tomb of a central figure for the faith.* It was also a stopping point on the Zagros Mountain Trail.

On Sinjar, almost one hundred thousand Yazidis were displaced when ISIS swept across the region. The militants branded them devil worshippers. This is a slur that has been used to justify oppression and aggression against the community for many centuries. Even now, in some places in Iraq, I heard it. In 2014, it contributed to genocide. Thousands of men were killed, and close to seven thousand women sold as sex slaves. Girls as young as nine and ten were sold at markets and were victims of the most atrocious sexual assaults, married off to fighters, and forced to convert to Islam. An entire generation of women were abused and traumatised. Many now lived in camps like Khanke, with thousands more still missing.

The camp was windswept and dusty, and the curve of white tents against a grey sky made it look as if their colour had drained into the lake beyond. We had an appointment to visit a young musician that Emily knew who ran a women's choir. Rana greeted us by the gate. She was twenty-two, with shoulder-length black hair, and she wore red lipstick that glistened when she smiled. In a friend's courtyard, on a tiled floor in the shade of a balcony, she took out a guitar and a circular *daf* drum. The *daf* was the size of a bicycle wheel with animal skin stretched tightly across it. On the inside, small metal rings linked together so they chimed when it

---

* Once, when I had asked about the importance of oral traditions for Yazidis, I was told: 'We carry the important things in our heart, because enemies can burn your books and destroy your shrine, but they can never take what's in your chest.'

was struck. The instrument was sacred, said Rana, and she cradled it on her lap like a child.

First, she sang a song that she'd written herself. Her voice was resonant, and long fingers picked out a simple guitar rhythm to keep time, nail polish dancing over the strings. The simple setting made me think of musicians I knew at home, in Ireland, sitting around in a huddle listening to someone share something new. She continued to smile as she sang, and the melody came loudly, confidently.

Then she picked up the *daf* and sang one part of a choir arrangement that she had worked up for her group. This sound was entirely new to me. She held the *daf* by the base in her left hand and the fingers of her right drummed one by one, galloping faster and faster to build to a crescendo. The left occasionally moved ever so slightly, sending the metal rings into a frenzy. The pitch of the rings reverberated in a kind of harmony with Rana's voice.

'Yazidis have always had a big focus on music,' she said when she paused. Her friend brought us all small, single-use plastic cups of water.

'There were many people who tried to erase our culture and music, and this is our way of keeping that alive. On Sinjar, my home, the worst thing imaginable happened. Music is part of us standing up again, protecting our heritage.'

She caressed the *daf* again, like a rider stroking a horse. Her nails tapped on the hard wooden rim. She wanted to tour, she said, to Erbil, and Sulaymaniyah, and to Europe. Music was her healing. It made her feel at peace, and she knew the same was true for others. The choir for women was to show what they were capable of. 'People here think they can take advantage of women,' she said. 'They say we're not smart. We're here to show them that nothing can stop us.'

Chapter Eleven

Tsunami

Day 17
Mosul Dam | Wana
River miles: 336

Access to the river was impossible again until close to Mosul city, and we left the Kurdistan Region in a rented minivan through a simple checkpoint. Bored Peshmerga scanned visas and waved us on. For the next few hundred miles, everywhere we would go had been held by ISIS at some point between 2014 and 2017. A few hundred yards further, we entered Federal Iraq by another checkpoint. This time, the armed men checking documents were members of the Popular Mobilisation Forces, or *Hashd al-Shaabi* in Arabic. This umbrella group of militias had been formed in 2014 in a desperate response to the speed with which ISIS was overrunning the country.* Since the announcement of Iraq's liberation from ISIS on 9 July 2017, in which the *Hashd al-Shaabi* played a significant role, the militias have become a permanent fixture in Iraqi security.

---

* After the fall of Mosul, the premiere Iraqi Shia cleric Ayatollah Ali Al-Sistani issued a *fatwa*, or religious decree, urging Iraqis to defend their cities and their country.

Exactly how that is defined continues to evolve. There would be a lot of *Hashd* ahead on the river. In our planning for the journey, we had received permissions that put us in contact with the Federal army and intelligence services, but the militias had a separate chain of command. At this first checkpoint, we waited with more than a little anxiety. At least Emily and I did. Claudio, as ever, was unfazed, and Salman told him off for trying to film the soldiers with a discreet action camera. Smiling, he put it away, and went back to sleep.

A slight soldier with a prodigiously straight moustache that seemed destined for his ears came to stand beside us. He wore brown fatigues and his tactical vest had various intimidating looking patches stitched onto it. One was the white skull from the Marvel comic and TV show *The Punisher*. This type of thing was familiar to anyone travelling in Iraq since 2014. The *Hashd* controlled many of the checkpoints in and out of Mosul and reaped the benefits of trade taxes and kickbacks. Some foot soldiers, like this man, seemed to model their appearance on Hollywood war films, or video game characters. But he smiled in a way that belied his apparel and showed us videos of his kids. This intimacy was also not unusual, and Emily and I had become used to approaching every military scenario by expecting the worst, hoping for the best and feeling confident that at the very least there would be tea.

A commander eventually waved us through, but within less than a mile we paused again. To reach Mosul Dam, a unit from the Iraqi Counter Terrorism Service, ICTS, insisted on escorting us. They travelled in two black Humvees, each with an armour-clad gunner peering out from under a helmet. We smiled and they grinned back, oblivious to the oddity of our convoy. One soldier at the checkpoint by the dam approached to ask where we were from. He spoke

English like an American and wore a heavy moustache, and I guessed he was around twenty-five years old. His shoulders were broad and he had a handshake like iron. These men from the ICTS were in shape. That was more than could be said for a lot of the other police and army that we'd met, many of whom rested potbellies on belts as proudly as they did their handguns.

The ICTS were famous in Iraq for their role in the Mosul war. They were known as 'the Golden Division' and held up as a model for bridging sectarian and ethnic divides in pursuit of safeguarding the nation. They were elite and rigorously trained. The Americans shared intelligence and equipped them like their own special forces. They had led from the front in Mosul and the surrounding towns and villages, and were often called in to support other police and army units under attack. The consistent fighting during that time, which was never the intention for a specialist force such as this, also meant they suffered heavy losses. As many as 40 per cent of Golden Division recruits were killed during the war.

The man with the American accent chatted easily with us, although he wouldn't share his real name. When Claudio moved to point a camera, he ducked away. 'We must always be in the shadows,' he said. He had a childish, cheek-pulling grin, and looked so much younger joking around with Emily and Salman. 'I'm a killer,' he told them, smiling, flexing his muscles.

'But you look like a baby!' Emily told him. He rolled his eyes.

'Okay, so I'm a baby killer!' he said, before realising the wordplay and slapping his forehead.

He was partway through helping Hana into the gunner's seat of the Humvee when his commanding officer arrived and put an end to the fun. This officer had one of

the largest chests I've ever seen and none of the baby killer's
boyish charm.

'You're ready,' he told us. 'Stay in your car and don't film
my men. If you do, I'll know.'

That was enough of a threat to keep even Claudio in line.

Three years after the impounding of the Atatürk Dam on
the Euphrates in 1983, Iraq put Mosul Dam into operation. It
was originally called 'Saddam Dam', planned in the 1970s in
reaction to Turkish and Syrian projects. At the time, Saddam
Hussein's government was replete with oil money and on a
spree of grand-scale infrastructure building. The Iran–Iraq
War put an end to that, but the Saddam Dam had begun
before the escalation, and construction continued even as
the economic and casualty toll grew higher along the border.

The chosen site was problematic from the outset. Salman
knew an engineer inside who'd agreed to talk, and we
followed Humvees towards the meeting point. Inside the
secure walls were communities for the many workers at the
dam and these mostly looked like any other small village.
One was home to thirty Christian families, and we drove
towards a metal cross and parked up on a wide street, lined
with cement houses and corrugated iron sheds. In the middle
a simple gate opened onto a square rose garden and, beyond,
a modest church.

The building itself was single-storey, plain and rectangular
with a glass door. A simple table with a white cloth functioned as an altar, and two wooden crosses hung on the wall
where they jostled for space with air-conditioning units. A
knee-high statue of the Virgin Mary, a cross-stitched scene
of the Last Supper and seven plastic roses completed the dec-
oration. Behind, rows of varnished pews tried to fill space.

Ibrahim, the engineer, stood in the doorway watching
us. He was Muslim, but had suggested we meet here. The

church was built in 1990, he said, taking his guiding duties seriously. Another man, who lived next door, had come to listen in, too. He sat alone at the back, heavy hands deposited on his knees.

It had all been razed by Da'esh, Ibrahim said.* He remembered when they arrived. It was 6 August 2014. 'They came in pickups. It was about two months after they'd taken over Mosul,' he recalled. He had fled, along with most other workers, and the dam was taken quickly.

'We knew they wouldn't let Christians live,' said the man on the pew. 'Anyone who stayed, hid. Our brothers kept them hidden.' He gestured to Ibrahim as representative of Muslims who protected them.

'The church was destroyed,' said Ibrahim. But when it was safe to return, the community rebuilt it first, and only then returned to reconstructing their homes. At first I had been struck by absence; the emptiness of a room built for dozens of families. But that underestimated the capacity of faith. To those who prayed here, it was overflowing with the holy spirit, miraculously recreated after the very hell of ISIS.

Someone on the street had brought a tray of rust-coloured tea for all the soldiers. Children arrived to watch, too, and Ibrahim picked pink daisies from the churchyard for Emily's and Hana's hair. The man who made tea waited until everyone had finished, then collected the glasses and returned home. Sometimes the intrinsic hospitality in Iraq could be overwhelming. It seemed to me the very opposite of ISIS, who persecuted others based on perceived differences. A much more enduring trait of the region was to reward strangers for those differences. So much of the population treated it as both a duty and a privilege to serve those

---

* Like everyone we met, he used the derogatory Arabic term for ISIS. I've generally used Da'esh instead of ISIS when the words are someone else's.

unknown to them. This is what Iraqi friends were getting at when they told me they were fed up with the media fixation on ISIS. To them, part of the healing process was returning the focus to these traditional ways of living together.

The Humvees groaned around tight bends as they climbed a switchback, and at a scrubby clearing we looked down over a rock wall, two miles long, branded in English with 'Mosul Dam Project'. This held back the almost four hundred billion cubic feet of water in the reservoir that we'd travelled across by boat. The dull, matt-coloured spillway for excess water had been unused for some time. A pool of deep, azure water had gathered underneath it, and the Tigris regained its lithe shape only when it snaked out to the south. Four chimneys at the far side of the wall marked the hydroelectric facility.

The dam served three purposes, said Ibrahim. It generated power, allowed for control of the water level to safeguard against flooding and provided water for irrigation projects in the surrounding countryside. Then he pointed to two areas at the bottom of the wall where white spurts of water foamed out. It was also leaking, he said, on a massive scale.

The problem was that, among the limestone and marl, it had been built on a foundation of karstified gypsum. Gypsum is soluble in water. The engineers at the time knew this, but were under pressure from Saddam Hussein to build, and deemed this area the least bad option. But ever since its impounding in 1986 it has been seeping, with sinkholes forming at the edges of the dam. The solution, said Ibrahim, has been what they call grouting, which involves pumping a concrete mixture into the holes which form under the dam. It's not possible to know what size the voids are until they're filled. Some are as big as a two-storey house.

To stop grouting would eventually lead to a failure of the dam. Inside the rock wall are sensors and warnings systems, but the risks are still terrifying. I asked Ibrahim what was at stake, and he said simply, 'I know you know what will happen. I don't want to say it.' He glanced nervously at the soldiers, but they were bored, and firing small rocks into the dam with a slingshot.

'Let me just say, at the minute, it's not sustainable. I'm an employee, so I can't say much more.'

Various reports made by American and European companies, as well as a public warning from the US Embassy in Iraq, suggest that roughly six million people are at threat if the dam was to breach. In a worst-case scenario, a tsunami wave eighty-five feet high would crash over the earth-fill embankment, reaching the city of Mosul in an hour and forty minutes. Anyone within a three and a half mile radius of the river would be washed away. Further south, the majority of Iraq's wheat fields would be flooded as the wave engulfed Shirqat, Tikrit and Samarra, before arriving sixteen-feet high in Baghdad within four days. Between half a million and a million and a half people could die, and many times more than that displaced. The major cities of the countries would be inundated, and the economy would surely collapse alongside the infrastructure for transportation, power and much else.

When I mentioned some of these projections to Ibrahim, he winced a little.

'I told you. I know you know,' he said. 'There are six million people at risk – it's not just me. The government knows that, but most experts tell us that it's been well done and is in a good way.'

In 2014, ISIS militants held the dam for nine or ten days. That posed the biggest threat to the dam's stability in its history. It is unclear how much, if any, permanent damage was done by the lack of grouting maintained during that period.

But now, said Ibrahim, they worked around the clock. Every other job at the dam got weekends off, but engineers like him in the pumping stations worked rotations to keep it moving 24/7. He pointed to one of the three stations. It was just below us, a scatter of grey steel and concrete. Ibrahim apologised that he couldn't say more. 'I'm glad you saw the church,' he said again. 'That's the most special place for me here.'

It was dams that had so far left the biggest impression on the shape of the Upper Tigris. Unlike other major infrastructure projects, dams affect a much larger area than just their nucleus. In Turkey the reservoirs of the Dicle and Kralkze and Ilısu had crept out insidiously, their tree-like branches reaching deep into the canyons and gulleys of the mountains. Millions had been displaced in the areas we had passed through; ecosystems were irreversibly damaged, heritage sites lost. Mosul Dam had the potential to fail rather suddenly, and it was entirely the product of ego and poor planning and a broken system.

Back at the checkpoint, the baby killer shook our hands. He squeezed Emily's extra hard, playfully, but it hurt her and he looked embarrassed. Another soldier, also young, complained to Hana that sitting in the turret hurt his back. His friends teased him, and they all grinned and jabbed at each other with soft punches. But when they'd gone, he told Hana that during the war in Mosul he'd spent fourteen hours a day inside the Humvee. The sore back hampered him when he played football or carried his son. So often these hardened fighters, to whom Iraq and the world owed such a great debt, felt like teenagers trying desperately to recapture some of their youth. It was probably impossible. They had seen so many lifetimes' worth of war and suffering, and yet they'd missed even more. There was a sense of lost time. For many military men in Iraq, they approached the autumn of their lives never knowing peace or stability. It was unclear how these soldiers would be rewarded in the long term.

We drove on to Wana, where ISIS fighters had first careered out on Toyota Hilux pickup trucks towards the dam. The village still bore the scars of the war openly, as if there might have been a skirmish that morning. Telegraph poles along the road had been snapped in two like pencils, top halves hanging uselessly by the shoulder of the road. In some places, new poles had been set, usually right alongside the damaged original. I had noticed across Iraq that there were rarely resources to remove defunct equipment, so, if something broke, it often sat rusting alongside its replacements. This was true for many things: telegraph poles, water pumps, vehicles, agricultural machinery. Even fences would sometimes exist in layers, with the newest at the front and rusted, flattened panels behind.

In Wana, many of the thick, flat cement roofs had cracked in the middle like wafers, crumbling at the sides. Iron girders twisted out. Other structures had been reduced to rubble, and these small pyramids of brick were almost everywhere. I had heard the battle for Mosul had left ten million tonnes of wreckage in the city alone. I had not found an estimate that included the surrounding areas, too.

Rehabilitation in the village had happened in stages. Generally, little money was available to those trying to reclaim their homes. Six million people were displaced in Iraq by ISIS, and in this area there had been heavy damage by coalition airstrikes attempting to remove militants holed up in civilian homes. This was the most common cause of the collapsed and distorted building. ISIS fighters often booby-trapped what was left before they finally fled too, so that family homes that weren't flattened might have had tripwires and IEDs strung throughout what remained.

The military did their best to remove these, but it was an enormous job. During the initial liberation effort, there was always pressure to move fast and show quick gains that politicians in Baghdad could shout about. Corners were cut, and extensive areas were simply left to be cleared another time. Sometimes families did it themselves. All of this meant that for those who did come back, sometimes two or three years after being forced into exile, they often had to rebuild a new home beside the wreckage or mined ruins of their old one.

We waited inside a grocery store for a boatman call Haji Ra'ed. The prefix *Haji* was attached to a man's name if he had made the pilgrimage to Mecca,* which Muslims are traditionally meant to do at least once. It was also used as a shorthand to suggest someone was old. Salman had taken to teasingly calling Claudio *Haji*, even though he was only fifty-eight, and about the least likely person I could imagine making a pilgrimage. The rest of us adopted it, and Iraqis who overheard found it hilarious. In Wana, the shopkeeper enjoyed it so much he called two friends to tell them.

We were eating reconstituted ice creams from a grumbling freezer when a major from the *mukhabarat* arrived in a white pickup and told us there'd been a security concern close by. Nothing to worry about, he said. Just some Da'esh. But we'd have stay in the minivan until Mosul. This was frustrating news. I had begun to wonder whether ISIS was a convenient excuse to monitor us, but it was something we couldn't argue against. There were still remnants of the group in the mountains here and to the south. They also had hideouts on islands in the Tigris. A UN report I saw said there were between eight and sixteen thousand ISIS fighters left across Iraq and Syria. The major first said he

---

* Women take the feminine form, *Hajijeh.*

had no idea if this was true. 'But I'm the smartest person you'll meet,' he confided, and winked at us. I suppose it was meant to be encouraging.

Chapter Twelve

The Boundary

Day 18
Kifrij | Sheikh Mohammed | Mosul
River miles: 368

The regional road along the river was pockmarked and among the early grasses of spring, toppled buildings lay like molehills. The major from the *mukhabarat* told us not to stray from the tarmac because of mines. There were no shepherds to be seen, which showed he was right to be wary. The war sat heavy upon these villages, and there were checkpoints every half-mile.

At a village called Sheikh Mohammed we were stopped. This checkpoint was controlled by a branch of the *Hashd al-Shaabi* called *Kataib Sayyid al-Shuhada*, KSS, or the Masters of the Martyrs Brigade. The KSS have close ties with the Islamic Revolutionary Guard Corps in Iran, and at times have been commanded and funded directly by Iran. They are a designated terrorist organisation by the US, and very vocal in their opposition to the American presence in Iraq.

It was unclear if they would let us through. It was also concerning to be in proximity to such groups. The *Hashd al-Shaabi* were diverse, but on the more worrying end of the spectrum for us were groups like KSS, who might see an incentive to kidnap us. It was unlikely, but I remembered a conversation with a security consultant in Erbil, who an NGO friend had kindly convinced to speak with me. His general advice about the journey was 'don't go', but the more specific part was, 'don't get kidnapped'. It would be unpleasant for my family, friends, and government, he said. I noted it would probably be unpleasant for me, too. 'If ISIS catches you, you're fucked,' he said. He liked the drama, but he was right. 'If it's a militia, like Hezbollah, it'll depend on the political landscape. The risk is pretty low, as long as you're not doing something stupid, but if things change, and they do all the time, then you might be worth something to them. So just stay clear. Or better, don't go.'

KSS had formed as an offshoot of Hezbollah. I wondered if trying to access the river here, with a cameraman, counted as stupid. I was happy to have the major from the *mukhabarat* with us, although I wasn't sure exactly how much he could help. The forces under Federal control were often at odds with these militia groups, who undermined the power and status of the national services.

With little else to do I tried to strike up a conversation with a militiaman. I had been learning Arabic sporadically for a couple of years, but made the mistake of jumping between dialects and teaching methods. As a result, I had a reasonable comprehension but I wasn't always good at articulating myself. My Iraqi dialect was weak, and I'd bought a phrase book from a shop in Erbil to help me acclimatise while on the trip. To my feeble ear, Iraqi sounded harder and more guttural than the Levantine words I'd learned.

The book claimed to be the 'essential language guide for contemporary Iraq' and on the cover noted how it would help with everyday conversational situations. The examples it offered were medical relief, security and reconstruction. Most of my passing conversations in Iraq were about weather and family and weekend plans, and I felt this 2004 edition might be a little outdated.

Once we had exchanged greetings, I flicked through and found the 'At a checkpoint' (*nuqat taftiish*) chapter, lodged between 'Demonstrations' (*Muzaahara*) and 'Searching a house' (*taftiish beit*). This seemed just what I was after. The first suggestion – Slow down! There's a checkpoint ahead! ('*la keifak, aku nuqtat tafiish giddamna*) – seemed redundant. The militiaman, cradling his gun, looked at me quizzically while I scanned the page. The next phrase – There are soldiers with weapons (*Aku junuud musallahiin*) – also seemed unhelpful. Instead, I bought time by telling him I was a writer from Ireland, and he nodded slowly. I flicked to the 'Searching a house' section and found the phrase:

'We are looking for: Ba'athists/looters/party members/ saboteurs/thieves/weapons' (*Ihna hnaa nfattish 'an: Ba'hiyyiin/ nahhaaba/hizbiyyiin/mukharribiin/haraamiyya/asliha*). This was new vocabulary, but still not ideal. Noting the accent and making a slight alternation for our circumstance, I carefully told him: we are looking for . . . *boats*.

He raised an eyebrow. I adapted 'Do you have any weapons here?' ('*dkum asliha hoon?*) and asked again about boats. 'No,' he said slowly, speaking as if I was dangerously stupid. 'There are no boats here,' he told me. 'No one can use the river.' All the book offered next was: 'Don't lie to us!' (*La-tkidhbuun 'leina!*) or 'We must take you for questioning' (*Lazzim naakh dhak wiyyaana lit-tahqiiq*), neither of which seemed like they'd go down well. I wondered what kind of off-the-shelf phrase book taught foreigners to yell at suspected Ba'athists and saboteurs. Instead I went back to what I knew well. 'Do you like football?'

For the next twenty minutes, we discussed the English Premier League. His favourite player was Lionel Messi, and he hoped that Messi would sign for a Premier League team. He thought he'd heard of my team, too, Aberdeen, because occasionally he looked at the Scottish results. But Scottish players were slower and sometimes fatter than English ones, he said. I wished I could disagree, and I also wished my Aberdonian grandparents were still alive so I could tell them about this conversation.

The soldier was from Nasiriyah, in the south, and had wanted to be a footballer when he grew up. He'd given up those dreams to rid his country of ISIS, he said. There was probably truth there, but his being in this area was complicated. Historically, many of the villages we were passing through had been populated by Sunni Arabs. There were also villages of ethnic Turkmen, who were predominantly Shia. These Turkmen were often descended from Ottoman migrants during the days of the empire, but some claimed lineage back to Turkic warriors employed by the Umayyad and Abbasid caliphs in the seventh and eighth centuries. The Turkmen of today had been targeted by ISIS. Turkmen villagers reported Sunni neighbours turning on them when ISIS arrived. Now many checkpoints in the area were manned by Shia Arabs from the south, as well as by Turkmen militia, and the Turkmen villages were heavily protected. The section of river we were trying to visit belonged to one of those.

Several men at this checkpoint were Turkmen, and since liberation from ISIS there had been many clashes between Kurdish Peshmerga, Shia Turkmen, Sunni Arabs and occasionally Shia Arabs from the south, too. It was an ethnic and religious melting pot, and in the climate that followed the war, old grudges were being played out in new dynamics. Now Turkmen militias were being accused of ethnic cleansing. The common response, alongside denial, was that their people had been victims of abuse from the Sunnis in the past. Now militia groups regularly arbitrarily detained men from the villages. Thousands had disappeared.

The KSS soldiers agreed to escort us to the river under the supervision of their commander. As with most military men that we met, he gave his *nom de guerre*: Abu Hajer. Seven of his men piled into an enormous green Ford F-350 pickup truck with a light machine gun mounted on the back, and told us to follow them to the Tigris. Shortly after we started driving, a small truck carrying gas canisters cut in between us, trying to overtake. The rear bumper, long and narrow, was painted with the sultry eyes of a woman peering out from under a *niqab*. It was a strange combination of lust and conservatism, and disconcerting to drive behind. The soldier manning the machine gun yelled at the truck driver, and soon we were back in convoy.

The Tigris was a mess when we reached it, and I could see why the soldiers were so dubious about us wanting to make a pilgrimage to the water. An excavator was digging out the riverbed, piling gravel on the bank nearest us. We walked towards it with the militiamen for company. All carried automatic weapons and wore battle vests with ammunition, medical kits and other supplies. Their uniforms matched, which suggested they were relatively well funded. One man had a helmet on with the visor down. Another wore brand new New Balance running shoes and stepped carefully over muddy puddles, tutting when colleagues splashed him.

The arm of the excavator plunged in and out of the water. In the distance, three more did the same. The sound of the diesel engine drilled into my skull, and it was impossible to hear anything else. When he saw us, the operator cut the power.

'How do you feel about the noise?' asked Emily.

'I have an iron head!' he laughed, tapping his forehead.

He worked here from 8 a.m. until 6 p.m., six days a week, he said. His boss lived in Mosul. I asked if there were any regulations here to protect the river, and he laughed again. No one cared, he said, and gestured towards the river. 'Does this look like something worth caring about?'

He was right. This was far from the mighty Tigris I had imagined. Even if we had been in Haji Ra'ed's boats, we wouldn't have made it through this gauntlet. 'Imagine what that means for the fish and other animals,' Salman said to me. There was a noticeable absence of anything alive here, other than men with guns and men with engines. A polluted haze hung over the water.

The excavator operator invited Emily to come up to the cab, and soon she was swinging the metallic arm wildly, sending *Hashd al-Shaabi* men diving this way and that. They found it hilarious, which was a relief. When she came down, the operator told us both quietly that actually the noise did really hurt, and he couldn't sleep some nights because of the headaches. Perhaps we had a tablet he could take?

The *Hashd* left us back at the checkpoint. Emily had told them that she and I had met while walking the Arba'een pilgrimage, from Najaf to Karbala. Now the man with the visor scribbled on a piece of paper. 'We have a *mokeb* there,' he told Emily, and he wrote down a number.*

The man had just come back from the shrines at Mashad, in Iran, and was very grateful for the opportunity. He reapplied some aftershave, then shook our hands and sent us off with Allah's blessing. The next time I had occasion to think

---

* During Arba'een, a *mokeb* is a hospitality tent, where volunteers prepare food and offer pilgrims a place to sleep. Each has a number and they were sometimes used to coordinate meeting friends and family in a crowd of millions.

of KSS was shortly after our journey finished, when they were suspected of firing rockets at Erbil airport.

The Tigris bent east in a final stuttering curve before working its way south to Mosul city. Most buildings close to the road had been obliterated by airstrikes. At village junctions, signs posted by the Mines Advisory Group showed pictures of types of ordinance left behind by ISIS, alongside information on what to do if they were discovered. In scrubland that rolled out to the horizon, tough, thorny grasses caught the ubiquitous black plastic bags that were given out in Iraqi markets up and down the country. On the thorns these fluttered in a light breeze so it looked like the plants were flowering, but also that those flowers mourned what had become of the land.

In a village called Kifrij, we met the Federal army. Perhaps keen to impress their superiority over the *Hashd*, they also brought a Ford F-350 truck in beige, plus two armoured Humvees. Occasionally I thought back to my original vision of this trip, in which a couple of friends and I paddled gently downstream on a raft. No wonder nobody took me seriously.

At least twenty soldiers tumbled out of the vehicles, much friendlier than the militia, and three men in long flowing *dishdashas* floated down a dusty track. One was the *mukhtar*, or mayor of the village, and another a religious sheikh. The third seemed a hype man for them both, and walked slightly ahead, heralding their arrival. 'Welcome!' he shouted. 'Sheikh Yasin welcomes you to Kifrij. May God's blessings be upon you. I have here also the *mukhtar*, who is happy to see you.' We exchanged lengthy greetings with the three men, and our group accepted hearty welcomes to Kifrij. As we walked along a mud embankment between two fields of potatoes, the *mukhtar* held my hand, interlocking our fingers.

This was common for men to do in parts of Iraq, and I was happy to be so well received.

The river was broader here and there were no gravel pits. Occasionally hawthorn and juniper grew by the water, but the Alpine and Mediterranean flora we had come to know was long gone. The geese had just left, flying north, and taken winter with them. At the bank yellow reeds grew higher than our heads, and among them were small, plump chukar partridges, little bellies puffed out in front. These rather odd little creatures were, in fact, the national bird of Iraq. In Kurdistan I saw them often up in the mountains. One of the local guides from the Zagros Mountain Trail used to make a living catching and selling them, until he became weighed down by guilt. He quit, but sometimes when we took a break from walking he'd watch videos of them on his phone, like a recovering alcoholic staring longingly through the window of a pub.

The village had always been peaceful, said the *mukhtar*. They were Sunni, and had good relationships with everyone else around. When ISIS arrived, jihadists swept across the river and, for a while, took control of the village. Eventually, the Kurdish Peshmerga pushed them back.

'Da'esh were afraid of the Peshmerga,' said the *mukhtar*, speaking to us while also fielding phone calls on an old, cracked handset. For the rest of the war, ISIS and the Peshmerga were both content to keep the Tigris as a border, right until the final offensive. But for the villagers, that meant being in a no man's land. The sheikh said they were grateful to the Peshmerga, and the village had stayed under Kurdish control until 2017.*

---

* In 2017 the regional government of Kurdistan held an independence referendum. At that time the Federal government, which refused to recognise the validity of the referendum, moved troops to take back Kirkuk and several other disputed areas.

'We're all Arabs here,' said the sheikh, 'but the Peshmerga liberated us, and we're grateful.'

The sun dropped low and the mud brick of the village glowed like bronze. Soldiers hurried us out, and we followed them back to the checkpoint. We drove past the site of another dam at Badush village. This one was half built, with brutalist concrete triangular blocks rising out of poisoned blue water. It had been planned to take the pressure off Mosul Dam but was abandoned in 1991 when Saddam Hussein began diverting funds elsewhere. It looked like a piece of modern art, almost comical amid the hills at twilight. Mosul soon appeared as a throng of traffic and blaring horns in the darkness. Then streetlights blinded us through the window, and ahead the growl of a city drifted in through open windows. We were a third of the way through our journey.

The great ziggurat at Assur, in central Iraq. Today it is eighty-five feet tall, but once stood at least twice as high. More than four thousand years old, it was part of a temple complex dedicated to the god Assur. Its six million mud-bricks were covered with sheets of iron and lead, and inlaid with crystals. Leon McCarron

Emily: We spent our days in the company of the military and intelligence units. Some relished the lifestyle and brotherhood; others would have rather taken different paths in life. I become especially close to one unit. Most of them dreamed of other careers as artists, designers and musicians. They loved the band Queen. Emily Garthwaite

A drone shot of the Tigris as it passes through Tikrit. On the near bank are the ruins of a complex of palaces built by Saddam Hussein. Once sprawling, luxurious and gaudy, they are mostly now destroyed after decades of conflict following Saddam's fall. Today the only inhabitants are members of the Hash'd al-Shaabi. Claudio von Planta

Emily: I think this is the most important photo I have taken. These are the last remnants of a red carpet leading up to Saddam Hussein's bedroom and living quarters in Tikrit. I took the time to walk up those stairs, kicking away bullet casings and cigarette butts. Through the window was a view of a mass grave dug by ISIS. Emily Garthwaite

Emily: The heat was biting that day, and Um Qusay was fasting. We had only a moment to take this photograph. Iraq's light is often unforgiving, so we stood under a tree and I photographed her through the leaves. She looked at me like this all day. Emily Garthwaite

The team that followed the river in Iraq, standing in front of the Malwiya minaret in Samarra. The distinctive minaret, alongside the Great Mosque of Samarra, was built eleven and a half centuries ago. Left to right: Salman, Leon, Claudio, Emily, Hana. Claudio von Planta

Emily: The farmer stood atop cracked earth beside the stagnant, toxic Diyala River. The air stank. I was convinced the river was all but dead, but then he pointed to the small heads of terrapins swimming near the banks. I was filled simultaneously with hope and heartache. Emily Garthwaite

Emily: I photographed Salman and Ali as they discovered the pipes churning out toxic wastewater. Whenever we encountered new ecological damage, Ali and Salman would both withdraw inwards. After years of advocating for the river in the face of so many challenges, each new experience like this seems to compound their suffering. I often wondered how they found the strength to keep going. Emily Garthwaite

The team walking towards Taq Kasra, the grand arch of the Persian city of Ctesiphon that existed somewhere between the third and sixth centuries AD. Much may still be under the ground, unexcavated. Claudio von Planta

Emily: When I first met Leon, his dream was to paddle down the Tigris, from source to sea. I told him that it sounded lovely, and that I would join him. In reality, I went down the river because I was in love, and where Leon was going I wanted to go too. It wasn't until near the end, here in Qurna, that I actually got to take a picture of him paddling. Emily Garthwaite

Emily: A feast of plenty in southern Iraq. In all my years in Iraq, I've never experienced such scrutiny with the hospitality. Each handful of rice I took had an audience of many. I politely declined to eat the grey meat coating the goats' heads. Hana told me it was a terrible day to be a vegetarian. Emily Garthwaite

Emily: Abu Haider started serenading his wife, Um Haider, as they drove their boats through the Central Marshes. Abu Haider is a famed Marsh Arab guide. His singing was certainly memorable. Emily Garthwaite

Our boat in the Hawizeh Marshes in southern Iraq. Beyond the expanse of open water, under the rising sun, is Iran. Leon McCarron

Emily pictured on the approach to Basra, with the rusting, decommissioned cranes that once hauled shipping containers in the background. Just north of here the Tigris is joined by the Euphrates to become the Shatt Al Arab as it journeys to the Persian Gulf. Leon McCarron

Emily: Ameer is one of my best friends. Whenever I visit Basra, I stay with him and his son Mo. They're like family to me. When we see each other now, we talk about how we can build a better world for Mo. Emily Garthwaite

At the mouth of the river, where the Shatt Al Arab spills out into the Persian Gulf. Shortly after this picture was taken, we all jumped into the water, being careful not to swim towards the far bank that is the border with Iran. Leon McCarron

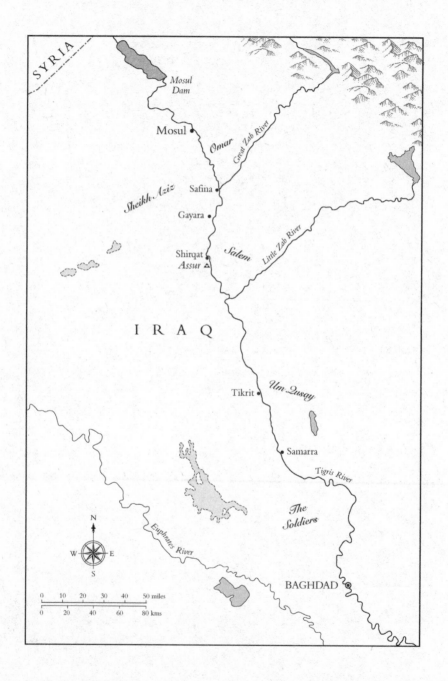

PART TWO

MIDDLE TIGRIS

MOSUL LAKE TO BAGHDAD

And it shall come to pass that all who look upon you
Will flee from you, and say,
'Nineveh is laid waste!
Who will bemoan her?'
Where shall I seek comforters for you?

Nahum, 3:7 (New King James Version)

Mosul is a splendid city, beautifully built; the climate is pleasant, the water healthy. Highly renowned, and of great antiquity, it is possessed of excellent markets and inns, and is inhabited by many personages of account, and learned men.

It has, besides, parks, specialities, excellent fruits, very fine baths, magnificent houses, and good meats: all in all the town is thriving.

Al-Muqaddasi, The Best Divisions for Knowledge of the Regions,
tenth century

Do not waste water, even on the banks of a flowing river.

Prophet Muhammad – Hadith No. 425, Quran

Chapter Thirteen

Hanging Gardens of Nineveh

Days 19–22
Mosul
River miles: 368

We rested up in the optimistically named Nineveh Palace Hotel. The rooms were as cheap as the curtains were threadbare, and what passed for a shower was a mouldy faucet plumbed into the wall at twice the normal height. We all loved it immediately, mostly because we could stay in one place for a few days in a row. The sparse, well-used furnishings meant we felt comfortable unloading everything onto the floor, and there was also no danger of staff coming to clean the rooms and interfering.

Close by was a restaurant that Salman liked, and we liked it too because the chef greeted us when we entered. He had an enormous stomach and wore a tight T-shirt that read: *'Dear God, if you can't help me be thin, at least help me make my friends fat'*. I told him it was funny. 'Oh, really?' he asked, smiling. 'Why? What does it say?' He presumed the English was meaningless, and its suitability for an overweight chef turned out to be mere chance.

In general, I'd noticed there was a large market in Iraq for cheap polyester T-shirts from China emblazoned with

English slogans. I enjoyed seeing these as I travelled, and they always provided a little levity. Some Iraqis knew the text was gibberish and didn't care, and others just never thought to ask. In Mosul I saw '*Only Battle Luck Rising*', '*Never take King time*' and '*American motor time good time*'. Some, like the chef's, were a little barbed, and others even became risqué. I wondered about the factories that produced them, and even more about who designed them. Was it an algorithm? And did anyone at any point in the process ever imagine the scenario that in Mosul in 2021 there would be a fourteen-year-old boy wearing a shirt that said, '*A blow job is better than no job*'?

Behind the restaurant were the old city walls of Nineveh. The high embankment sprouted tufts of grass in early spring, and in the evening young couples sat overlooking the motorway below. Some of the area inside the walls, once part of the city of antiquity, was now used to graze sheep. When we took a walk there we found a few illegal homes built haphazardly beside a grove of palm trees, and a camel that looked even more out of place than we did.

Nineveh was, for almost a hundred years, from 705 to 612 BC, the capital of an empire at the height of its power. It was one corner of the triangular-shaped Assyrian heartland, reaching out across the plains to Arbela – modern-day Erbil – and south along the Tigris to Assur,* home of the god that lent his name to the empire. Nineveh was close to a natural ford on the river and functioned as a terminus point for overland trade routes that came through the mountains. It gave access to Anatolia and on to the Mediterranean, while Assur on the western bank opened up the Euphrates Valley and the Babylonian areas to the south. Arbela helped control the Zagros and routes towards Iran.

---

* To Assyrians, the sovereign power was always their deity, and the mortal leader merely a chosen representative.

Early European explorers began digging at Nineveh in the middle of the nineteenth century, when the area was under the yoke of the Ottoman Empire and permits obtained with relative ease in Istanbul. One of these adventurers, an Englishman named Austen Henry Layard, was to make monumental discoveries both here and nearby. He worked with his assistant and protégé Hormuzd Rassam, an ethnic Assyrian from Mosul. At Nineveh they excavated a mound called Kuyunjik, close to the banks of the Tigris, and there discovered the palaces of the kings.

The findings of Layard and Rassam shed light on the fate of the Assyrians, which had intrigued writers since classical times. The fascination was perhaps because of Nineveh's mention in the Bible, in which it was destroyed by the wrath of God. Layard and Rassam, usurping earlier French attempts at excavations, quickly began pulling out immense human-headed winged bulls, or lamassu, carved from gypsum. These were the guards of Assyrian palaces. As they went deeper, they found more sculpted bulls and lion sphinxes, as well as gold, jewellery, and, as Layard wrote: 'halls, chambers, passages, whose walls, almost without exception, had been panelled with slabs of sculptured alabaster, recording the triumph and the great deeds of the Assyrian kings.'

In the palaces of Sennacherib, 705–681 BC, were depictions of his monumental building projects, including the transport of the lamassu. In the palace of the slightly later Ashurbanipal, 668–c.631 BC, there were mostly stories of war, featuring full-bearded men with bulging biceps, and of the king's own exploits as a lion hunter. The lion was the seen as the most dangerous animal during Assyrian times, and the reliefs commissioned by Ashurbanipal show in gory detail the death of the lions, bleeding out and punctured by spears, sometimes rearing up to rip at their attackers.*

Brutality was central to most of the scenes. The Assyrians have gained a reputation as vicious, ruthless warriors, and it's easy to see why. Neo-Assyrian kings were relentless campaigners. Anyone in their path was forced to swear an oath to the god Assur. If they did not, they were usually killed, often in pretty nasty ways. One account of Ashurnasirpal's conquests reads like this:

> I took the city by assault and I skinned alive all the leaders of the rebels and I built a pillar which I covered with their skin. Some of them I walled up, some I impaled on the pillar, and others I tied to stakes around the pillar. Some royal officers had rebelled and I dismembered them.

One might think that scribes carving into clay would be economical with descriptions, but when it came to torture and death, no verbosity was spared. The account continues:

> I burned many captives alive and many I took alive. I cut off their noses, their ears, their fingers; I put out their eyes. I made one heap of the living and another of heads, and I hung their heads from tree branches around the city. I burned the young men and women alive.
>
> I bricked up twenty men in a wall of the palace alive. All the other survivors I left to die of thirst in the desert.

The message, here as elsewhere, was clear – don't mess with the Assyrians. Subsequent kings left writings in a

---

* The last lion in Iraq was killed in 1926.

similar vein.* The Assyrians had the largest army in the world – which by the mid-ninth century was employed year round – and the greatest territory. Conquered populations were often moved elsewhere within the empire to quell unrest.

It's possible that this bloodlust is overstated somewhat, because the commissions of the kings were made to project power and may not have fully reflected reality. What is clear, though, is that those kings were equally adept administrators of their territories. From the cuneiform texts pulled up alongside the reliefs, we learn how the Assyrians managed their armies, cities and subjects. In translation we can read of the luxury that Ashurbanipal lived in, and in particular the wonders of his royal gardens. These were irrigated by the Tigris, and said to have extended some thirty miles into the mountains. Plants from all over the empire were gathered to be planted there. One Mosul-based historian who I'd spoken to before the journey was working on proving his theory that the fabled Hanging Gardens of Babylon were actually in Nineveh.

Layard's most enduring discovery at Nineveh was a collection of many thousands of clay tablets in different languages covering a variety of topics. Three years later, after Layard had left the Middle East, Rassam found more on the opposite side of the mound, and today the collection of twenty-six thousand tablets is known as the Library of Ashurbanipal. Ashurbanipal collected these texts from across the Assyrian Empire, in Akkadian, Sumerian and Babylonian, which detail myths and histories of the Mesopotamian civilisations, as well as religious and divination texts and legal documents.

---

* I was somewhat disappointed to learn that Shalmaneser III, who had begun to feel like a companion of sorts as we moved downstream past his various reliefs, had once used the bodies of enemies to make a bridge over the Orontes River in Syria.

His library has been an integral source in the understanding of ancient Mesopotamia.

Ashurbanipal was not the last king, but he was certainly the last of the great kings. By 614 BC, thirteen years after his death, the Medes had destroyed Assur and two years later, along with the Babylonians and Scythians, they brought down Nineveh. Layard and Rassam's explorations showed the world the workings of the Assyrian Empire at its height, right until its dramatic, rapid fall.

Layard wrote a bestseller, and his name became synonymous with the findings at Nineveh and nearby Nimrud. But, as the Assyriologist Gwendolyn Leick puts it, 'Nineveh was found too early, before archaeology had been invented.'

It was not until the turn of the century, via German pioneers, that a more scientific approach to archaeology reached the Middle East. Layard's excavations, like all during the Victorian era, destroyed unknowable amounts of the heritage of sites he visited. He notes in his writings how painted frescoes and ivories and other fragments of the buried city 'crumbled to dust' or 'fell to pieces' as he worked.

He and his contemporaries crated up and shipped pieces as fast as they could pull them from the ground. Naturally they were taken by kelek on the Tigris to Basra, then transferred to ships. One even ended up at Canford, a public school in the south of England, where Emily had boarded. Layard brought back a series of alabaster friezes from Nimrud in gratitude to a patron who had paid for him to write a book, and over time the room where the artefacts were displayed became the school tuck shop. In 1992, John Russell of Columbia University identified the stone slab beside the dart board as an original frieze from the throne room of King Ashurbanipal. It was sold for six million pounds at auction and helped build a new sports facility and theatre at the school. Layard, like his contemporaries, took the heritage of the world as his own, and used it as currency to build his profile and refill his coffers. There was an understanding of the value of the pieces he pulled and stacked, but certainly not of the cost of his actions.

The activity of the excavators in the mid- and late nineteenth century eventually led to rampant looting across Mesopotamia. By the end of the nineteenth century, Orientalists dispatched to the region spent as much time buying artefacts from antiquities dealers in Baghdad as they did working the sites. Some explorers of this era clearly had great academic knowledge and understanding of where they stood, but little awareness of the historical travesty that they were creating. Excavations were driven by desire for fame, or riches, or as a tool in the ongoing quest for control of the Orient. Post-colonial history has not been much kinder to the antiquities of Iraq. At Nineveh, like so many other sites, much still lies beneath the ground, and more still perhaps has been destroyed or looted.

On the morning of Easter Sunday we walked the walls of ancient Nineveh, passing a neo-Assyrian arch through which the modern road had been run until ISIS blew up a sizeable chunk of it. The jihadists sold oil on the street beside the wall, a police officer told us, and happily destroyed any part of the site that seemed too offensive to them.

We followed the line of the wall to a quiet, residential neighbourhood and there found Father Ra'ed, whose charisma was matched in magnitude only by his moustache. The church was called Bishara. Like many Christian sites in Mosul it had been destroyed by ISIS, and two armed guards patted us down before we could enter a newly rebuilt and spotless courtyard. A couple of hundred other worshippers were already inside, and we joined the last stragglers to take our seats.

The new interior was long, low and air-conditioned. Criss-crossed ropes decorated the roof, like a large-gauge fishing net, and the harsh light of morning was filtered through modern stained glass. At the front, Father Ra'ed welcomed us all, joking with those in the front row and sending his voice booming down the aisle. Altar boys busied themselves with incense and prayers.

'Let us praise the Lord!' said Father Ra'ed in Arabic. He told the story of Jesus at Easter, sparing no detail. Every few moments, he'd make eye contact with someone. He was very good at his job. Then he spoke about peace-building, and the similarities between Christians and Muslims. They had similar scripture and sayings; both practised non-violence. If someone hits you once, let them hit you again. He smiled a winning smile. When he was done, and a few hymns were sung, Basim Al-Abasa, a Muslim member of the Iraqi parliament, stood up from the VIP area at the front. She thanked him for protecting the Christian heritage of their city, and asked him to always remember that Da'esh were not representative of Muslims in Mosul.

'We should protect all of you. What you said about peace-building was powerful.'

Later, the conversation continued in a side room, with Father Ra'ed sitting at one end and various important figures squeezed onto sofas along the sides. In the empty space in the middle, young men rushed around handing out coffee and wrapped sweets. One had a T-shirt that read, '*If you want happy, be better*'. The governor of Ninewa, Najim Al-Jibouri, was there, in an immaculately cut blue suit, and he strode across the room to join the priest at the front. A man beside me leaned in close.

'He's tough,' he said of Al-Jibouri. 'They call him the American general, you know. Because he has so much special American training.'

He glanced around, then told me a story. In 2005, when the town of Tell Afar was overrun by Al-Qaeda, they desperately needed a good general, he said. The last two who took the job were killed on the day they started. Then Al-Jibouri came in. 'He cleaned it up. No problem. He's a tough guy.'

The man squeezed my shoulder tightly.

A brief series of formal, complimentary speeches began, with Basim Al-Abasa continuing where she left off. She spoke beautifully and seemed genuinely touched by the experience of the service. I was, too. This was my first visit to Mosul, and what struck me most was how normal the mass had seemed. Watching the battle for Mosul from Erbil left me emotionally distanced from it. How could I ever understand what had happened there? Moslawi friends in Erbil had told me that one consequence of the ISIS occupation that really frustrated them was that many people, foreign media in particular, seemed to forget the city's long history. Everything became simply about war and destruction.

'Mosul has always been coveted,' one friend told me. 'It has a long history of people trying to fight for it.' The Tigris cuts through the heart of the modern city, but rather than splinter it in two, it was historically the very reason for Mosul's strategic importance. It's Arabic name, *Al-Mawsiil*, means the 'linking point' and may refer to the city as a crossroads, linking Diyarbakır with Baghdad and Basra by river, and connecting caravan routes from the west and east. It was from Mosul that a Seljuk governor led the Muslim retaliation to the Christian crusades in 1104. Under the Ottomans, it became the capital of a province that covered all of modern-day northern Iraq.

It was also known as a city of diverse religions and ethnicities, with Kurds, Assyrians and Turkmen living alongside Arabs, and Christians, Jews, Yazidis and others sharing the space with Muslims. It was about this thread that Basim Al-Abasa and others spoke most eloquently.

'Your speech was very important for social cohesion,' she said. 'This is what we need.'

A Sunni tribal leader stood up. 'Christians are central to this city. They are the life of it. *Insha'allah*, they will return.'

Father Ra'ed shook his hand and responded. 'This is my life, from the beginning. There is true love and true peace here from our Muslim brothers and sisters and has been ever since I was welcomed back three years ago.'

In 2003, before the American invasion, Father Ra'ed told us later, there were forty-five thousand Christians in Mosul. Slowly that number fell as a result of conflict and disenfranchisement until 2014, and ISIS, when every single Christian left. Of the Syriac Catholic churches in the city only Bishara is still operational, and Father Ra'ed believed just seventy Christian families have returned.[*]

He was a busy man, with a lot of hands to shake, and I moved back to the baklava and coffee tables. There had been three other foreigners present, visiting with a French Christian NGO, and I saw them ushered out by a no-nonsense nun. Then she caught sight of me. 'Allez!' she said, whipping a wad of papers at my shoulder. I tried to protest but she wouldn't hear of it, in English, French or Arabic. She herded me to the main gate, where a minibus waited, whipping the papers with venom every time I turned to speak. We reached the doors of the bus and I was at a complete loss. It seemed ill-advised to pick a fight with a nun on Easter at

---

[*] In April 2022 a second Syriac Catholic Church, Mar Tuma, also reopened in Mosul.

a church in Mosul. But I also didn't want to have to join a French NGO. Inside the bus, French staff stared at me. 'I'm sorry,' I mumbled. 'The nun brought me here.' They nodded sympathetically, and I realised they, too, must have been browbeaten into places they weren't meant to be. It took four of us, but eventually she accepted I wasn't French, and didn't need to go with them. She gave me a final whip with the papers and sent me scurrying back to my friends.

Chapter Fourteen

They Came by the Bridges

Day 23
Mosul Old City
River miles: 368

I was carrying two tourist guidebooks to Iraq from different periods. The first was published in 1973, just five years after the Ba'ath party took control of the country in a coup, but before Saddam Hussein emerged as the party strongman. In 1958, revolution had overthrown the British-installed monarchy and, after the ensuing turmoil, the early 1970s was a time of relative stability when Iraq's economy was booming. Diplomatic relations were forged with the US and the USSR, and there was hope that prosperity for Iraqis was finally there to stay. The 1973 guide, published by the Tourism and Resorts Department, reflects that. It's beautifully illustrated with scenic and cultural highlights and has sponsored full-page advertisements from tour companies and airlines. There is an entire section on resorts in the country, focusing especially on 'Habbaniya: Iraq's Sunny Beach', and suggestions for water skiing, boat races and swimming contests.*

In the book we see how Iraq was still experimenting

---

* Midway between Fallujah and Ramadi, the lake's fortunes have since fallen.

with how to present itself to the world. The text is humble, descriptions brief. 'General Information' notes that illiteracy has been cut from 80 to 50 per cent in thirteen years, and that the country has placed a new emphasis on revolutionising agriculture. The country's theatre movement is 'vigorous', ceramics undergoing a 'revival', and one of the few more expressive pieces of writing puts these advancements down to 'the black gold, which suddenly gushed out of its generous soil'. But the book's primary purpose is functional: a first attempt at promoting to the world what the authors saw as an emerging country of interest.

The 1982 book – *Iraq: A Tourist Guide* – is thicker and more robustly printed, and on the first page is a smiling portrait of 'Field Marshal Saddam Hussein, Hero of National Liberation', who had formally taken power three years earlier. The second page is the score for the new National Anthem, entitled 'Land of the Two Rivers'. There is a celebratory tone from the outset:

> Progressive, revolutionary and triumphant Iraq finds inspiration in its great history and the nation's magnificent heritage as it develops itself and builds up a better life for its people consonant with their yearning and aspiration. The Tourist guide in your hands, dear reader, provides a true picture of some aspects of this country's great heritage and civilisations, its arts and sciences, its growth and progress, together with its tremendously varied landscape ... We trust it will introduce you to the Iraq of yesterday's civilisations and the Iraq of today's progress and triumph.

Ten years earlier, there was no mention of government or politics. Now Chapter 1 opens with 'Form of Government', 'Internal Policy', 'The National Assembly' and 'Foreign

Policy' (including abidance by the UN charter, good relations with neighbouring countries), followed by a history of revolutions, the country's development plans and investment ideas, and a summary of important ministries. Only then are we led on to Baghdad. In the glossy photographs of men and women poolside that follow, or the suggestions of casinos and bars to visit on tours around the country, we get an insight into the Iraq that Saddam Hussein wanted to project.

Both books were also a window into the country before it became a battleground for US troops, before the Baghdad Museum was looted and sectarianism writ large, and long before ISIS. Despite the obvious limitations of state-sponsored guidebooks, it was also clear that forty to fifty years ago, Iraq's heritage and environment were much healthier than they are today. 'Mosul is called the *city of two springs*', according to the 1982 *Tourist Guide*, 'because spring and summer are so alike. It played a leading role in Arab wars of conquest and became a city of great importance.' Pictures of tourists in the Old City illustrated the words. To flick through the book in Mosul in 2021 was a melancholy experience.

Moslawis orientate themselves according to the direction of the Tigris, much like how the Upper Nile is south of, and geographically under, the Lower. In Mosul the east bank is left, and the west right. To us, heading downstream, this made sense. Bishara church was on the left bank, as was the Nineveh Palace Hotel and the ancient city. After the fall of Nineveh the city of Mosul rose on the right side to become the pre-eminent power. The Old City was the heart of this.

Mosul had fallen to ISIS on 10 June 2014, when a few thousand militants swept into the city and chased out terrified and poorly prepared Iraqi army units. For two years, jihadists occupied the city and prepared their defences for a siege they knew would come. On 16 October 2016, more than one hundred thousand fighters began the effort to liberate Mosul. The coalition was drawn across religious and ethnic backgrounds in Iraq, along with significant international support.

By this point, ISIS had been mostly defeated in the rest of Iraq and the areas surrounding Mosul. The coalition closed in on somewhere between three and twelve thousand fighters still left in the city. They began with a series of airstrikes to take out the bridges over the Tigris. Then ground troops moved in, cautiously clearing city blocks, destroying ISIS positions and leaving clear paths for civilians to escape. It was incredibly complex, and progress was slower than expected. ISIS were deeply embedded and had left IEDs and roadblocks strewn across the city.

It took until the end of January 2017 before the liberation of eastern Mosul. The army secured access to the destroyed bridges, and began the offensive across the river against the last stronghold. ISIS used the walled Old City as the keep at the heart of their fortress, and withdrew for a final battle there in which the only certainty was obliteration. It was July before the Old City was liberated, by which point between 65 and 80 per cent of it was destroyed.

Eastern Mosul had felt much like any other Iraqi city. Shops were open, and the smells of kebab and shawarma filled the street. But at the riverbank, everything changed. The walls of the Old City, which had once pressed out towards the river, were a mess of rubble. Some exterior structures remained partially standing, jagged and useless. Others looked like they had melted, or that someone had simply dropped thousands of tonnes of bricks from a great height. It was total annihilation. Hundreds of years of history lay as scattered debris, or buried alongside bodies and bombs, and no birds flew overhead, as if even they knew there was nothing there for them.

We crossed on one of only two operational bridges. It was patched with great iron girders while the original struts dangled below. For a whole day we walked through the Old City, around winding alleyways that once led to and from bazaars, and past iron doors that enclosed the homes of Moslawis, now dead or gone. Many of those doorways had ornate, carved stone façades. Most were festooned with bullet holes alongside the snaking detail. Here and there businesses had started back up, often in a single rebuilt unit amid a ruined street.*

Everyone had a story, and every story was enough to break your heart. The owner of a falafel shop lost all his children in an airstrike. A man smoking a cigarette by a blackened wall said it used to be a public hammam where 130 civilians were executed in a day by ISIS. Next to him was a wreckage that was once a medical clinic. An unexploded rocket had been found that morning, so red and white tape was put up to deter anyone from wandering in.

A policeman, off-duty, took us to the Armenian Orthodox Church of Our Lady, which had been used by the *Hisbah*, or morality police of ISIS. The man with us said he'd been inside to the nave shortly after liberation, and there were bodies everywhere and piles of cash stacked against the walls. In the dungeon, the *Hisbah* tortured people who broke their rules. The policeman said people could end up there for not sporting a full beard, or wearing trousers the wrong length.

'ISIS thought the churches wouldn't be hit, so they used them for important activities,' he told us. Their reckoning was only partially correct. In the same area, three other large

---

* 138,000 houses were destroyed in the city during the war, 53,000 of which were on the west side.

churches, each of a different denomination, once marked the centre of Mosul's Christian heritage. Now the four buildings shared a collective ruin, the roof of Al Tahira blown off to expose blue, bullet-hole-ridden walls to the sky. Ornate interiors were scorched with soot and blown through with holes the size of cars. Underfoot we saw the fine, sketched lines of Syriac script on cracked alabaster.

I wondered how Salman and Hana felt walking among such recent ruination in their country. But both had seen it before, and so much more. They were of a generation of Iraqis that had known conflict since their earliest memories: born during the war with Iran, high school during the Gulf War, university under sanctions, fledgling careers amid American occupation and sectarian fighting. Whenever I asked either of them about what we were witnessing in the Old City, they shrugged. It was terrible, horrific, beyond comprehension. But it was also the latest in a series of atrocities that had occurred throughout their lives. They did not have the luxury of innocence or shock.

On our way back to the river, Emily and Hana drifted off into a market. When they came dashing back across the busy road, a watching traffic cop rotated his hand, as if turning over an hourglass, in the classic gesticulation of *what on earth?* They had a met a woman called Khitam, Emily said, and she pulled us back into the road so we could meet her, too.

Khitam was in her late forties and wore faded jeans and a brown overcoat with a polka-dot scarf. Her hair was cropped and the creases on her face led to a large scar on her right cheek. She was the only woman we could see in the market, and the only one we'd seen anywhere on the right side who didn't cover her hair. She shook our hands, and as vendors gathered around to listen she shooed them away. They sniggered and whispered. 'I don't give a shit,' she said, smiling, kicking out at one in a way that felt only partly playful. She was from the city, she said, but left during the war. Did we want to see her old house?

The walk only took a few minutes and re-entered wasted residential streets. In Khitam's area, there were no longer any roads, no returnees, no rehabilitation, and the debris had collected in high mounds. Occasionally rockets had blown out the ground level, too, revealing basements or making craters. To walk meant bridging the gaps over these voids, then climbing over the next hillock of brick. It reminded me of trekking high on an unstable mountainside more than any urban experience.

As we drew close to Khitam's home, a Humvee packed with soldiers reared up over the rubble. Another followed, and two pickup trucks behind.

A soldier shouted at us to leave, but Khitam tutted.

'I lived here,' she called back, over her shoulder. 'These are my guests. Who the hell are you?'

We kept walking. The Humvee reversed back down the mound, and the commander called Salman over. They were *Hashd*, shooting a film, he said. It was about how they got rid of ISIS. That was surely much more important than whatever we were doing. Behind him, a soldier in full battle gear checked his beard in a pocket mirror. Salman stood his ground, until eventually a shy man with a small video camera raised his hand at the back. He was the director. They needed this area for a scene, he pleaded. Please could we be quick? Many of them had to get back to other work when their heroic battle scene was over.

Khitam's house had been hit by an airstrike, and the structure remained like a skeleton. The strata of her past life were blown open. She showed us what had once been her bedroom, another for her four children, and the kitchen. Then she sketched in the air where stairs were, and where she hung pots and pans and kept spices.

She had been married twice. Both husbands died in wars. The only person she could ever depend on was herself. She spoke freely. She liked women and occasionally dressed as a man. 'I don't care about gender,' she told Emily. Before the war, she was an assistant to a human rights lawyer during the day and borrowed her brothers' car at night. 'I'd drive it as a taxi, but only pick up women. I'd wear a *dishdasha* and a hat like a man, and I'd make sure I got them home safely.'

When ISIS arrived, they came by the bridges. Convoys of trucks streamed across, flags flying. Khitam didn't know what to expect. On the first day, nothing happened. On the second, the same.

'On the third I heard they were taking pretty women and selling them in Syria,' she said. 'And it got worse after that.' Black flags appeared above the bazaar, and food ran low. She heard stories of neighbours going missing. Her family hunkered down inside their home. Her husband, who was Kurdish, was killed, and she was left to protect the children and home. They ate grass that grew outside the doorway and, when water stopped running from the taps, they went to the river. 'Everyone was there,' she said, drinking from the Tigris, scooping handfuls desperately into their mouths and filling pans, always watching to see who was around. 'The Tigris kept us alive,' she said, though often they would get sick too from the pollution.

In the end, she fled when ISIS came to take her children. A large man with a red beard knocked on the door one day and demanded she hand them over. There was a scuffle and she was injured, but she protected the threshold. She knew he'd return, so the next morning she walked with her children to the outskirts of the city. 'I had to get out. My youngest was five months old, and I'd already run out of milk. If I could, I mixed dry bread and fruit and water to make a paste, and I fed her that. I didn't know if she'd make it.'

Eventually Khitam reached Kirkuk, and on to Sulaymaniyah. She had only come back to Mosul after liberation to speak to the courts about documents for her children, because when they fled it was with almost nothing. Now she leaned on the doorway of her old home as she spoke, occasionally pointing to something remembered. Eventually, Salman told us we had to leave, so that the *Hashd* could have their turn at filming. During our interview, he said, a couple of local men had also sidled up to him to whisper that Khitam was crazy.

She was not crazy. Even before the war she was different from most people, she told us, and men were afraid of that because they saw it as a challenge to their authority. Now, after decades of being told she was crazy, she had added the theatre that the men craved to prove their point. She hissed at someone who approached her, yelled an expletive at another. She was performing for them, because it was easier than simply being herself in a place that didn't have room for that. Even after a war, when so much had been lost, the patriarchal system had held firm. I wondered whether one day she'd become what they'd always accused her of.

Chapter Fifteen

The River Sheikh

Day 24
Mosul
River miles: 375

On our last evening in Mosul, Salman and I walked by the river and talked about hope. He said he was always optimistic, despite everything. 'You have to be, or you could never do the work I do,' he said. Even in all of this, I asked. He was sure. It was a chance to create something better, to rise like a phoenix from the ashes.

The sun set behind the scars of the city, and Salman and a team of thirty young Moslawis organised a litter pickup by the Tigris. For a couple of hours they worked on the riverbank, pulling plastic and bottles from stubby, sooty grass. The Tigris rushed by, dark cobalt in fading light. A little further upstream was a natural sulphur spring. At one point, this must have been a lovely place to relax, and bathe, and gaze back on the city.

Humat Dijlah had their work cut out. The Tigris was suffocating. The pipes we had seen were dumping untreated wastewater, loaded with nitrogen and phosphorus, directly into the river. That wastewater came from factories, hospitals and other industrial complexes. Some was initially discharged into septic tanks, but most went to sewers that led to the river.

The excess of nutrients caused an acceleration in algae and aquatic plant growth, eutrophication, which then produced toxins that affected water quality for fish life, birds and anyone who drank it. At its worst, it killed the fish and poisoned the water. The Tigris was already carrying the wastewater from towns further north, as well as the Kurdish cities of Duhok and Zakho, whose effluent ended up in the river via tributaries. Hazardous waste from industry around the area of the abandoned Badush Dam project, tonnes of buffalo shit each day, and raw materials from the sand and gravel mining ended up there, too. Agricultural run-off, which should have been carried out to the desert, also now came to the river. Salman's team had worked on itemising these pollutants, but the task had become much more challenging since ISIS. The Tigris had become a receptacle for all the garbage in the northern part of the drainage basin.

Bands of young men sat on the ramparts of the twelfth-century Bashtapia Castle, smoking water pipes under crenellations and a reddening sky. 'Why bother?' one shouted down to the litter-pickers. 'Haven't you seen the other shit going on here?' But Salman spoke patiently, kindly, about the future. If Mosul was going to rebuild, he said, it needed to do so in an environmentally sustainable way, or there would be more devastation. The volunteers nodded along, and one exchanged numbers with the heckler. Then a family came down from the castle, too, to ask if their young daughters could help.

We woke early to find a boat that could take us out of Mosul. In the arrowed light of dawn a series of small Kia-brand trucks trundled over the patched-up bridges to the central fish market, their flatbeds converted into pools for flailing carp. A small generator worked a pump that kept the water fresh, and the excess poured out of the tailgate, blotching the pavement like a trail of breadcrumbs.

Fish were heaved into jaundiced bathtubs, or old refrigerators set on their side. A ribbed roof kept off the sun, but not the flies. In the middle was a fire pit where fresh carp would be grilled for *masgouf*, the national Iraqi cuisine, in which the fish is killed, spatchcocked, roasted, seasoned and served, all within a half-hour, bathtub-to-table. Large men in welly boots and tracksuit trousers waddled between duties, shouting for tea from a scurrying young man with a silver tray, while the sun crept above the corrugations.

A sinewy boatman with a firm chin offered to help us. He introduced himself as Omar, and his camouflage T-shirt read '*West Virginia*'. Omar was quiet, but had an assurance that I liked. He spoke straight, and I trusted him. There was a military man who controlled access to the Tigris, said Omar, and he worked from a floating office in the north of the city. That was who we needed to see. Everyone called him the River Sheikh.

Omar brought us to the pontoon in his boat, and there we settled on a sofa with a sheepskin rug. Behind, a mannequin in full diving gear watched over us. Further along the river I saw the well-known *gazino* cafés. They reached out into the river on stilted struts, and at night families came to sit and smoke shisha by the Tigris. Occasionally, speedboats in party lights would take punters out and do donuts in the current for a few thousand dinars.

An hour passed, and then another. Claudio fell asleep and snored, and a young soldier brought us sweet tea every time we looked like we might protest. The river rushed by. Finally, upright and broad-shouldered, with a huge handlebar moustache, the man we were waiting for arrived, like Lemmy from Motörhead lost in Mosul. His uniform was pristine, and his chest showed medals but bulged like it wanted more. 'I'm the River Sheikh,' he said, matter-of-factly. 'I hope you're not French.'

He sat with us and smoked a cigarette, then offered the packet around. Emily took one, and he raised an eyebrow. He looked at me.

'Better that women only smoke in private here in Iraq, like here. She shouldn't do that in public. People might get a wrong idea of her.' Emily cleared her throat.

'I know,' she said. 'I can also make up my own mind.'

'As you like,' he said, now setting his eyes on her. 'But it's not like being a man. As a man you can run naked through the city, and no one cares.'

I asked about his moustache, hoping that some gentle compliments might soften his demeanour, though I was not sure there was softness in this man. 'The shape and the size comes from my life,' he said. 'I became a commando diver in '93. I've been a military diver ever since. If you live a life like I have, you earn this moustache.'

He had been trained by Americans, and they were the only foreigners he trusted. The Americans had given him six small boats, RIBs, and two big ones. The French, on the other hand, promised to help but brought nothing. All his three decades of service had seen war, or sanctions, or both. He'd learned to make do, he said, and gestured to the wet rooms on shore. There were half a dozen divers in there who did all the work for the whole Ninewa Governorate. That was a few hundred miles of river. There was no extra pay, and they bought their own equipment.

'Bodies,' he said when I asked what he dived for. 'Only bodies.'

As if on cue, to impress upon us the darkness of this work, a soldier rushed over. He saluted the River Sheikh, delivered a message, then dashed off again.

'Two bodies just now,' he said to us. 'Up near Sheikh Mahmoud. I'm dispatching a team.'

Behind us, four men suited up. One, Abu Shaqriya, had been doing the job for twenty-eight years. They were gone within a few moments, borrowing Omar's boat and taking another, back upstream towards the northern area where we'd had trouble accessing the river.

'I'll give you what you need,' said the River Sheikh, looking back at us, and lighting another cigarette. 'The Tigris has a lot to tell. But you're going to need two reliable boats, and two boatmen.' He complained one last time about the duplicitous French and their empty promises. I thanked him and he said again, 'I'm the authority on this river. You go with permission granted by me. But you'll have to rely on the drivers. They'll deal with any problems.'

Implicit in this was the threats further south – *Hashd*, ISIS, and what else? I asked who he recommended. He jerked his neck towards Omar and, for a moment, I was suspicious. We had found Omar almost at random by asking around the Old City. Or so I thought. But perhaps, just as we'd found the River Sheikh, our stream of questions about boats had led us naturally to Omar, too.

'He's one of us,' the River Sheikh said, and Omar nodded with a quiet, sullen movement.

'He's done more for us than almost anyone. Has he told you about the ferry?'

On 21 March 2019, a passenger ferry sank on the Tigris. It was carrying families to an island fun park where many were celebrating Kurdish New Year. It was desperately overloaded, with more than two hundred passengers on a deck rated for fifty, and capsized in the strong current. Most on board couldn't swim. In the end, 103 died. Omar had saved thirty-three lives in his fishing boat, repeatedly heaving them onboard and then to shore. Now, as he told the story quietly, he said that after a while he pulled up only bodies. He lost count of how many dead he carried to shore to lie with the living.

'Omar will go with you,' said the River Sheikh again.

We had to wait for Omar's boat to be returned by the divers. When they arrived, two men carried the limp body of a fifteen-year-old, whose limbs spilled heavily out of their awkward grasp. The other divers brought his brother, just twelve, wrapped in a blanket, so only his yellowing face and curled toes could be seen. Their uncle, a short, rural man in a pale *dishdasha*, shocked to silence, waited to be helped to shore. His legs were failing him.

The boys had been swimming and were pulled into a strong current. The uncle couldn't swim and was helpless to save them. They were the fourth and fifth drownings that week. 'Mostly it's the villages, like these two,' said the River Sheikh. 'We try to raise awareness, but often the kids go in with adults who aren't strong swimmers. They have no idea how dangerous the river can be, or that one day can be stronger than the last.'

The two boys lay there, lifeless and soft, a little foam at the corners of their mouth. I had never seen dead children before, and I understood now when people said bodies could seem peaceful, as if at rest. But that seemed more likely to be my way of trying to process such a tragedy. The divers formed a circle, raised their palms to the sky and the River Sheikh recited the *Fatiha*, the first chapter of the Quran.* He signed a form, and the bodies were lifted into bags.

---

* This is an obligatory prayer in death.

I looked at Emily, who was pale, and Hana, who stood silently by Salman. 'It's not my first time,' said Hana, when I asked if everyone was okay, and I understood her. Hana, full of sweetness and light, who rescued small birds with disabilities, had told me once about walking to school as a child in Baghdad and seeing two dogs sniffing at a human body by a railway track. Another time she saw a young boy from her school killed in crossfire on the street. One of Salman's best friends had been killed by Al-Qaeda for being Yazidi, and after 2003 he'd had to leave his family home and move in with an uncle in a safer part of town. Hana had spoken of a generation of lost children. When she experienced death now, in all its raw, limp horror, she processed it differently than I ever could.

We left with Omar, and as he manoeuvred out from the bank, the River Sheikh called to us. 'By the way,' he shouted, 'the divers were shot at by the *Hashd* at Sheikh Mahmoud.' That was the area where we had met the Kataib *Sayyid al-Shuhada*, and the men who shot at the rescue boat were probably the same soldiers. 'They think the river is theirs. They said we should have told them we were coming. So be careful.'

Chapter Sixteen

Warning Shot

Day 25
Lazagh | Hawsalat
River miles: 383

One of Salman's *Humat Dijlah* colleagues in Mosul, Bassam, said he heard us talking about the *Hashd al-Shaabi* a lot in English. Salman suggested we have a code word so that those overhearing us didn't get suspicious. We all spent a moment in deep thought. Then, with great clarity, Bassam looked at the fruit stall across the road. 'Watermelon!' he said. It was brilliant. It made us smile and held no sensible connection to the militias. From now on, we would speak about watermelon checkpoints, watermelon boats, watermelon permissions and risks associated with the watermelons.

The Tigris would take us south-east until the confluence with the Great Zab, some sixty miles from Mosul. Then it makes a gentle arc ever so slightly west until, after roughly the same distance again, the Lesser Zab also washes in. These tributaries were the next waypoints in my mind, and there were villages on both banks. Despite the dangers that lay ahead, I was confident that with Omar's company and the River Sheikh's support, we'd be able to navigate any problems that came our way. I was wrong, though I didn't yet know it.

We set out in two boats. The second was under the control of Ridvan, an ex-bodybuilder in a skin-tight base layer. He and Omar fished together. Ridvan smiled a lot, even when he peeled up his shirt to show where an ISIS sniper had shot him through the neck. 'Bastards,' he laughed. He had been selling cigarettes and other contraband goods during that time. Then he fought with the army against them. The wound didn't seem to bother him much, and he was happy to show it off.

With them was a third man, Daham, whose presence was a surprise. He was in his early twenties, though seemed younger, and did not possess the athletic confidence of Omar and Ridvan. Where they were experienced and savvy, he was happily naïve and clueless. He wore a Chelsea FC tracksuit and slouched up front on Ridvan's boat. Before we'd even left the city, one of his flip-flops fell into the river and, trying to retrieve it, he'd fallen in, too. Ridvan circled for a while before heaving him in with one arm.

'That's why we keep him around,' laughed Omar. 'He's funny.' It was the first time I'd seen him smile.

The mood, mostly because of Daham, was light, and we sped under the city bridges, steering around the tentacled girders. Daham composed a song about our journey, and each verse concluded with a plea to take him back to England. 'If you like it so much here,' he sang, with only the faintest hint of a tune, 'then I'll go to London, and you can live with my mother.' He blamed us for his broken phone, which he hadn't been able to turn on since he'd fallen in. 'But don't worry about a new phone now,' another verse went, 'because I'll be given one in London when you take me back.'

At its widest in Mosul the Tigris was two hundred yards,

dark and deep, but at the city limits, it was squeezed into a
gap of less than ten yards. Mostly it was because of the extensive excavation of the riverbed, which turned the Tigris into
an open quarry, and islands of aggregate were piled high from
the centre all the way to the bank. White bubbles and swirls
showed unknown disturbances below the surface. Pumping
stations also encroached periodically, their engines groaning
as they pulled water out to send to the villages. Our boats
slowed to a crawl to edge through what was left. The chug of
our outboard engines filled the space so that the sigh of the
river's movement was lost until we cut the motor and drifted
alongside the flotsam.

Despite these interferences, there was more life than we
had seen in a long time. Cormorants perched on rocks in
the eddies and, as the city withered, the banks grew wilder.
Water buffalo in the distance wallowed clumsily in the
shallows. On the artificial peninsulas created by the quar-
ries lounged clusters of young men, not dissimilar from the
buffalo, occasionally rolling into the water to cool off. They
smoked water pipes and, even at the early hour, we saw tall
cans of beer poorly hidden under discarded T-shirts. We were
at a glorious point of poise in the season, with the shivering
cold of winter gone and the desperate heat of summer yet
to arrive. For a fine, fleeting moment, a perfect temperature
hung with the light clouds above us. And, perhaps for the
first time on the entire journey, I felt a sense of flowing on
the river, like how it might once have been.

Two checkpoints under the control of the *Hashd* – the
watermelon people – waved us over to talk, but Omar had
already called them. Once Omar identified us, the soldiers
returned to their cramped, unpainted brick sentry points to
watch over the empty river. After the second checkpoint,
we stopped to speak with five men making charcoal. They

had brought tree branches on the back of a pickup and now stacked the wood in cross-hatched layers. Then it was covered in sand, with a hole left open on one side to let oxygen in. It took three days before the charcoal was ready, and then they'd bag it up and sell each for four thousand dinars, or just under three dollars.

It was hot, tiring work, and the men were caked in black soot. They did it because there was nothing else. The youngest and most talkative of the group asked about our journey. For our own security, we told no one we met that we were following the river to Basra. It would not be hard to ambush us if so desired. Instead, we told him we were just travelling for a day or two on the Tigris around Mosul to learn about life along the river.

'You want to hear a story about life on the river?' he asked. 'How about the story where the Iraqi military doesn't pay on time, so we go elsewhere?' He had fought ISIS with the army, but after the war he only received two months' pay out of twelve. It wasn't enough to live on, and he had a baby, so he joined the *Hashd* a year ago. They paid, he said, much better than the army, and on time.

As we moved south, we saw more brick lookouts, partially veiled by the reeds that now grew ten feet high on the left bank. The villages had receded, pushed back over time by flood waters when the river used to break its banks. Once those inundations were common because of rainfall and snowmelt at the headwaters. That was never predictable, and there were no great lakes to moderate the flow, as on the Nile. But since control of the river had tightened with the dams, and a changing climate meant less rain and snow in the north, this sweep of Tigris was wide, and low, and the spaces in between the banks and settlements became uninhabited marsh.

At another soft bend, the river divided to isolate a large island, and in the distance was a soldier on the gravel bank. He was apoplectic with rage, stamping on the ground and waving his AK-47 wildly in the air. His voice sped across the water, but the words were lost in the emotion. When we came alongside, I saw he was young, twenty at most. His face was red, and there was fear there alongside anger. 'I was about to shoot you,' he wheezed. Omar tried to clasp his arm, but he thrust it away. 'Who are you? Are you completely stupid? I was about to pull the trigger. I only stopped because I saw there were two women on board.'

Omar looked at him and said nothing. The man fought to control his breathing, not yet relieved that he had not killed us, though that would come. 'Say something, by God! I was going to kill you first,' he pointed at Omar, 'and then you.' He nodded to Ridvan. 'And once the drivers were gone, I'd wait to see what happened with the rest of you.'

Omar stood with him for a while and explained. The soldier knew nothing of us. He was from a branch of the military, not the watermelons, we learned. No one had called him. He held up a decrepit Nokia. The continuation of our journey, I realised, and indeed our lives, came down to this phone. We had been incredibly lucky. All along the river there would be soldiers tasked with watching the water, and unless they received a call on some ancient handset, they'd shoot. What if their phone was off? Or out of battery, or broken? I had no idea.

There were ISIS in this area, the man said. Until recently, they had been living on the island we had just passed. They came once from the reeds, by boat. A colleague had killed a boatful. The soldier was still shaking, and it was clear we would not pass his post. Omar told him we'd be back, and I got in his boat with Emily and Salman. We swung out into the current, arcing around to the other bank to turn upstream. As we did, the air beside us cracked, and the water around the stern fractured like glass. We sunk instinctively into the boat, but there was nowhere to go. 'Hey!' Omar shouted. 'We're turning around!' The young soldier held his weapon, still pointed at us. Perhaps he thought we planned to speed off past him. It must have been a warning shot, I mumbled to Emily, because surely he wouldn't miss from there, though it was scant consolation to any of us.

Omar and Ridvan nestled the boats in the reeds upstream and made calls. Our team were okay. Claudio found it mildly amusing. 'These guys are crazy,' he said, laughing. After a couple of hours, which we passed in contemplative silence, Omar said it was sorted. We returned cautiously, nosing the boats towards the bend. When we passed, the soldier stayed in his concrete box, and did not wave. He also didn't shoot.

I was no longer convinced our journey was feasible. How much of the river would be like this, with checkpoints every half-mile? The right bank rose to a small cliff, thirty feet high, and the grass on the top diffused afternoon light. ISIS once had tunnels and caves inside, said Omar. The water was deep here. In Mosul, ISIS often dumped bodies of those they'd killed in the Tigris, and the bloated corpses usually washed up here. 'It's safe now,' he said. 'Don't worry.'

'How do you know?' I asked Omar. We were out of second chances.

'I'm a soldier, too,' he said. He left it at that and brushed off additional questions.

I understood now why the Tigris was empty of traffic. Anyone using it would be shot immediately. As a technique to stop ISIS moving by river, it was effective, but it also deprived the Tigris of being any use as a waterway. When we began, I wondered whether the river still connected places and people, as it had done in antiquity and under the Ottomans, or whether it was a tool for division. The fiefdoms and fragmented sectors of control that we'd witnessed so far gave a straightforward answer. The words of my Moslawi friend in Erbil still rang in my ears. 'Everyone just talks about ISIS now when they think of Mosul.' I wanted to busy my mind with something else, anything else, but how could I? How could anyone? Millennia of history had not disappeared, but it was hard to think of archaeological sites when there were armed men at every turn of the river. Even Salman had stopped pointed out the sewage pipes.

Omar pressed us into the reeds again to make another call, and Daham tried a song, but his heart wasn't in it. Next on the list was the watermelon commander in the area. The nickname seemed facile in such serious surroundings, even in use among ourselves, but we stuck to our rules and half whispered it in an apologetic tone when we discussed whether the watermelon people might shoot us.

Salman translated Omar's conversation with the commander. 'The watermelon guy is saying: "Let me tell you something",' said Salman. '"You don't come into my area and then call me when you're already here. Leave my territory immediately and then ask permission."' Omar tutted. 'That's unnecessary,' he said, and began to introduce us.

'You have no respect,' the commander said, cutting him off, his voice distorted now through Omar's speaker. 'Get out, right now. If you come any closer, we'll kill you all.'

The commander would not budge, and the day was late. We went back to a safer area and camped on high ground. The next day, a friend of Ridvan's took our boats on the back of his truck around the *Hashd* area, and also past another island that Omar told us might still have ISIS on it. The idea that there could be terrorists surviving on an island of marshes less than a mile long, with all these soldiers around, seemed baffling. Always present, too, was the realisation that for us this was a one-time journey. For those living in Ninewa, along the river, it was the reality of life, and access to the river for commerce, transport or recreation was strictly controlled or banned entirely.

The villages along this stretch of river were some of the saddest I had seen anywhere in Iraq. In one, Lazagh, every resident from the ninety houses had been displaced by ISIS. Most homes had been destroyed, and just two were rebuilt. In one, on a rise overlooking the river, we drank tea in a neat rose garden. The owner said Lazagh was included on a map that Henry Layard commissioned of this area in the 1850s. 'We're still proud of that,' he said. 'Our village is on a map in the British Museum. Look around. Think what this would have been like before all of this.'

In Hawsalat the *mukhtar* said, 'The Tigris is the spirit of this land, and it feeds everything we do.' But now the water was so low in summer that it wasn't enough to irrigate their wheat and barley fields. Normally, spring brought increased flow and would carry all the garbage and pollution downstream. Initially, this meant rubbish from Mosul washing through, but then for a couple of months they had clean water again, like in his youth. That hadn't happened for the last few years, and he mentioned climate change. It was the first time that someone we'd met had attributed the increasing temperatures and periods of drought to this. Most blamed Turkey and the Iraqi government, in that order.

The *mukhtar* was a kind man, but drooping eyes showed a deep fatigue. He counted on his fingers six young men from the village who were in prison for joining ISIS. Emily met a fifteen-year-old girl in a home close to the river whose husband had also been arrested nine months earlier for suspected ISIS sympathies. Her first husband had been killed in 2017, fighting as a jihadist. The girl was alone, with a child to care for, and terrified.

ISIS recruited heavily from villages such as these. There were surely many more that had fought and died for the caliphate than the handful the *mukhtar* counted out. We were at the northern edge of an area often called the Sunni Triangle and for decades farmers in these areas had suffered from poor policy and increasing environmental challenges. Under Saddam Hussein, agriculture was sidelined in favour of oil production and wars. After the American invasion, villages throughout Ninewa and Salahaddin Governorate to the south were targeted by jihadists. One man in this area said his cousins had joined Al-Qaeda to fight the Americans. They eventually went to Fallujah to train, and he lost contact during the years of sectarian conflict that followed. A few years later, he recalled, other strangers started arriving in the villages. Some were from Mosul, some from Fallujah, others he wasn't sure. They'd make speeches and offer something better.

'They'd remind me how hard it was to make a living now, and that the government was the reason,' he said. By 2012 and 2013, the visits were more frequent, and farmers were promised money if they joined the fight. It wasn't exactly clear what that was, at least in these villages, but the visitors leaned heavily on religious dogma and pre-existing rifts. He remembers them telling him, 'Your water is running out, and the government doesn't care for you.' The government at the time was led by Nouri Al-Maliki, whose divisive, sectarian-based policies alienated and disenfranchised Sunnis throughout this region. 'I never trusted them,' the man said of the bearded visitors, but plenty of others did. The offers of money and fighting back against a hostile government found many takers.

Those who ignored the advances in these rural Sunni areas, like the villagers we spoke with, had lived through a hellish, helpless time. Some of their young men left to fight for ISIS in Mosul. Most who did died there, and others ended up in prison. Villages were destroyed in the fighting, and homes that were razed were left booby-trapped for the oncoming coalition forces. Both here and further south, we heard stories of villagers being kidnapped during the war and taken to Mosul to be used as human shields.

Sometimes, when residents did eventually return, their neighbours were hostile. And so those farmers in Ninewa who rejected ISIS were left with drought, crop failure, destruction of their property and distrust from all those around them. ISIS, it was clear, too, had not gone away, and in other areas were targeting tribal leaders and *mukhtars* who had resisted them, carrying out kidnapping, assassinations, and burning more land. The man we met in Hawsalat, who would not share his name, was sick and couldn't work. Instead his six children, aged between nine and sixteen, worked all day making charcoal. 'What can I do?' he said. 'The government had ignored us. We're left here to die of thirst and hunger.' He thought for a moment and added, perhaps to be clear, 'But what the others did, with Da'esh, that was against God, and their punishment will be greater still.'

Chapter Seventeen

The Great Zab

Day 26
Hamam Al-Alil | Saff Al-Tutt
River miles: 426

There was once a time when the town of Hamam Al-Alil was known throughout Iraq as a place of healing. Its name, meaning 'baths of the sick', came from the sulphur springs. But although the marble massage tables of the hammams are still there, the town has met the same fate as the rest of Ninewa Governorate. It was occupied by ISIS for two and a half years, and still existed in that same liminal space as other places we'd seen: liberated, but not safe. Destroyed, not yet rebuilt.

This was Omar's home, and he lived with his wife and kids in a small house close to the river. His fishing boat was moored outside. The rest of the town looked like one might expect after recent history. Debris flanked the road, buildings still wore the shock of rocket attacks and throngs of men, young and old, lingered in doorways with little else to do. There were few women to be seen, except in a playground of plastic slides and rusty swings where they gathered with their sons and daughters, pushing them back and forth and huddling together to talk.

We slept in a semi-abandoned water treatment plant. Whatever bureaucracy, or bureaucrat, had once required an office here was gone, and we erected our tents in the bare rooms to keep the mosquitos at bay. There was a small shower room, but the flickering light inside attracted so many thousands of bugs that it was worth staying dirty. The building was inside a compound, and a dozen soldiers in full battle gear wandered in and out of the gate. We were forbidden from leaving unless Omar was with us.

A colonel named Arian briefly visited. When he stepped out of his Humvee, every soldier snapped to attention. Colonel Arian carried an iPad instead of a gun and nodded at my notebook. 'Write these down,' he said, and reeled off a batch of coordinates. On his screen he showed me a military map of his sector of control, with pin drops where there were active concerns. A thick finger pointed at an island to the south.

'That's the main one. You can't go past that by river. You'll have to go around. After that, it should be okay, but your guys will have to keep in contact with mine.'

He nodded and left, and the soldiers relaxed. One, sensing an opportunity to talk, came over sheepishly. He had an enormous moustache that curled out from his face and must have taken hours to oil and groom. As Arian had, he thrust a screen at me, but this one had no military coordinates. Instead, the soldier wanted to show me a video of himself, in Baghdad, at a conference for men with exciting facial hair. 'It's called the Crocodile Club,' he said, a little bashfully. 'You'll find us on Facebook. We do community service and help the poor, and then we meet up and compare our moustaches.'

We took a walk along the river to see the sulphur springs, and the Crocodile Club soldier tagged along behind Omar, twirling his moustache and caressing his Kalashnikov. I wished I cared about something as much as he loved those two things. We reached a large circular pool, twenty feet at its widest, which bubbled with creamy green, geothermal waters. Skinny kids with deep tans dive-bombed off the low brick wall that surrounded it. One reached into a cavity in the rock and showed me how he pulled tar off the inside of the spring. To my surprise, he popped it into his mouth. 'Chewing gum!' he said.

He offered me some, which was not appealing, not least because of how many sweaty bodies were being cleansed with the same water that washed the tar. Also, it was tar. But the kid chastised me, and his friends joined in, so I reached my hand into the narrow hole in the rock and pulled out a lump. I opened my mouth and chewed, and it was mostly tasteless but with a hint of salt. I pushed the thought of sweat from my mind. The boy waited for a thorough review, and when it never came he took out his own gum, borrowed a lighter from Salman and lit it. My only consolation was that I was glad not to have known it was flammable before I tried it, but I quietly spat out my own wad and moved on.

Close by, a pipe emerged from the ground to fill a small pool, dug out of rock, and two shirtless men sat applying mud liberally. 'The minerals here will cure just about anything,' said one of them, coating his pale chest. 'My son has a skin condition on his foot. I bring him here once a month to keep it clean.'

I joined them, encouraged that my body was no more a stranger to the sun than theirs. One of them helped me spread mud across my back. They were from Mosul and came here as often as they could. There were real hammams in the town, they said, where you could get a massage for a dollar. But these two Moslawis preferred the open-air experience.

There was something wonderfully simple about the spring, where status and other markings were discarded along with clothes. Everyone wallowed as equals. 'This'll heal anything,' the first man said again. 'I'll live longer!' He laughed, but his friend shook his head with a grin. 'It'll help, but only God knows what might happen. An idiot like you could walk up to the road and get hit by a car. These minerals won't help that.'

Omar walked with us, quietly, patiently. I felt he was honest whenever I asked him anything, but he was so frugal with his words that it was hard to know much about him at all. Just before we left the springs, he saw a friend called Muhannad, and the two shared a quick embrace. Muhannad was young and angular, with a scar low on his cheek, and lived in Hamam Al-Alil with his family. They exchanged a catechism of polite greetings, and I waited to move on. But suddenly Omar spoke up. 'Muhannad's father was a real hero,' he said.

Muhannad's father Ahmed was a farmer, no different from most until ISIS arrived. He hated them from the start and was determined to do all he could to unsettle them. When they turned a building by the river into a media centre to produce propaganda, he went in the night and set it alight. The ISIS cadres were furious, but Ahmed was a skilled agitator, and the deed could not be traced. Then he began to help others escape. Men with military, police or government connections were targeted, so he smuggled them out, across the Tigris, to Kurdish-held areas to the east. Once he was arrested by the Peshmerga, who suspected him of being a terrorist, but after twenty days they let him go. He returned to work immediately, driving a Toyota Hilux pickup to the riverbank in darkness, then rowing the silent, shaking bodies of the fleeing across the water. He helped around 450 soldiers, said Muhannad. 'One of the first was Omar.'

Omar looked at his feet. I'd had no idea. Eventually, he spoke.

'I fought ISIS everywhere as a soldier. I did a lot of work against them, not just fighting. Ahmed saved my life. There was a price on my head, and they were after me every day.'

Omar mentioned a brother now. It was the first we'd heard of that, too, but it seemed like this was the time and place where he was happy to share, and his story trickled out, still sparse, succinct, but every word heavy and considered.

His brother was also a fisherman. They learned from their father, and worked the Tigris up and down Ninewa, often competing to see who could catch more carp in an afternoon. The brothers shared a room at the family home until they got married, and after that their families spent free time together. They joined the army together, gathered information on ISIS and fought them in battle.

But his brother, a year younger, was not as lucky as Omar. He was tracked down, taken to the river and executed, hands behind his back, T-shirt pulled over his head. He'd been shot five times and dumped in the Tigris with two others. Omar fished out the body downstream. He also lost his cousin, one of three hundred bodies found at the local agricultural college, which ISIS had used first as a training ground, then as a mass grave. We'd passed it earlier that day, flattened to the ground, still filled with IEDs and perhaps more bodies.

Muhannad's father, Omar's saviour, also died. He eventually went into hiding, moving along the river in the reeds, but someone in a village gave him away and six trucks filled with ISIS fighters surrounded him. He killed seven of them before they got him, people say. His body was hung in the town square, eyes and digits removed. 'There was no one like him,' said Muhannad. Omar nodded once more.

We put in the boats at a bridge below Hamam Al-Alil. When the bridges were blown up in Mosul, this crossing was the closest for city travellers to use for over a year. It took an hour or more, often on dirt tracks, to drive from the Old City to eastern Mosul via the detour. Now two city bridges were operational, and the governor, Najim Al-Jibouri, had said a third was imminent. I thought again about the task facing him. Ninewa was the second most populous governorate in the country and had suffered the most damage in the war. It depended on agriculture and, although early in the season, it was shaping up to be a dry year.* The months when heavy rains were expected had passed. Farmers we'd met already complained of insufficient water for the wheat harvest. The level of the reservoir had dropped at Mosul, too, and Salman said wells in villages away from the river were drying up. All signs pointed towards a severe drought later in the year, and disastrous failure for the wheat and barley harvest.

Al-Jibouri had a bumper reputation as the major general at the heart of the battle against ISIS, and he was popular. He had spoken regularly about the next, most important, phase – rebuilding – and progress was being made. The iconic Al-Nuri Mosque was being rehabilitated by UNESCO, with funds from the UAE, and other organisations and local residents were working as best they could to piece back their homes. But from what I had seen of the rest of the province, the job was almost inconceivably large. In interviews, Al-Jibouri had spoken about Ninewa being forgotten about by Baghdad, and having to attempt the impossible of reconstruction with limited resources. That seemed like it could be true.

We crossed a small set of rapids on the river, which Ridvan said were natural. A historian in Mosul had told me that actually this was an Assyrian weir, created to manage the flow of the Tigris south of Nineveh. That seemed much more

---

* Later we'd learn it was the second driest in forty years.

believable, because the rocks ran in a straight line directly across the width of the river. Not far away, on the east bank, was the site of Nimrud, once called Kalhu and a capital of the Assyrian Empire, after Assur and before Nineveh. There Layard had unearthed vast relief slabs from the eighth and ninth centuries BC, great winged bulls that now stand sentinel in the British Museum, and thousands of smaller sculptures, tablets and ivories. His findings proved the location of a city previously only known to the west in the Bible.

The rocks of the weir breached the surface, just, and curled waves formed on the downstream side. There was a single way through, around the back of a small island, but even then the boats would have to run through a short whitewater section. This was another reason for the success of the *keleks* in the past. Their shallow draft would have had no trouble with the rocks, and, had we been on one, we could have drifted straight over. Instead, Ridvan disconnected the outboard motor, pulled the oars from their fastenings and rowed into the tongue of the current. I sat in the back with Emily, enjoying the adventure, slightly regretful that I could not row myself. In the past, I had spent quite a lot of time paddling kayaks and canoes in rough water, and it was strange to be a passenger. But Ridvan was strong and confident, and we bounced through easily. Omar's boat followed.

The historian had said this could have been one of a handful of fords where Assyrian armies crossed the river. Farther along, at a village called Saff Al-Tutt, there was a beautiful rock harbour with half a dozen fishing boats tied to a wooden dock. It was natural, too, said Ridvan, but I wondered whether the curves of the limestone hadn't been carved by hand. The steps up to the village certainly had, and it seemed possible they could be Assyrian. Everyone we met insisted it was just the shape of the rock. We didn't linger long, because in the shallow water by the jetty where we tied the boat was a large rocket, three feet long and unexploded. It had been there for five years, said a group of boys, and wasn't a problem because everyone who came knew not to hit it with their engines.

The right bank climbed higher, crumbling caves yawning out of mud, and the Tigris ran clearer, softer. At times we could see algae below, swaying with the undercurrent, young fish darting between the blades. On the left, the land flattened out. Then, with a sudden burst, the Great Zab River rushed in. Its journey had begun near Lake Van in Turkey, east of the source of the Tigris. There are still no major dams on the Great Zab, which makes it unique among the tributaries of the Tigris, and much of its 250-mile lifespan is spent coursing through tortuous mountains in the Kurdistan Region of Iraq. I knew the river well. Many of the trails that I had scouted there ran close to it. I had walked above and alongside, swam in all four seasons, camped by its banks. In one of the valleys on which the Great Zab is etched, on the flank of Bradost Mountain, is Shanidar cave where ten Neanderthal remains have been found. These date from sixty-five to thirty-five thousand years ago and are testament to the immense depth of history in the Zagros and Taurus mountains. It was from areas like Bradost, twelve millennia ago, that troglodytes slowly moved out to the foothills and plains and began experimenting with settled agriculture.

The Great Zab surged in at a perpendicular angle, tanned and creamy in contrast to the azure Tigris. Its colour came from the fields of Kurdistan, where farmers would pump excess water from the fields, sweeping pesticides, topsoil and anything else that moved out into the river. The Zab was significantly more powerful, too, driving our river towards the sheer bank. We followed this line of fluvial mixing, like milk settling in coffee, and for the first time were not alone. Four or five fishing boats, similar to ours, passed by on small circuits of the confluence area, and by a gravel shore another seven were berthed with fishing nets prepared in the bow.

This was Safina village, and as the last light of day leaked out over the western ridge, we set a fire and prepared our tents. Ridvan and Daham cast their net close to the far shore and pulled in two large carp. The smaller one they threw back, and somehow in the process Daham fell in again. Omar gutted the fish, carved them longways through the middle and impaled each on a roasting rack made of sharpened sticks. Daham stood shivering by the fire. He had only one set of clothes, so I lent him a jacket which he pulled round him dramatically. 'Great trip this is,' he complained, knowing it would bring a laugh. 'We get shot at, nearly hit a rocket, I get frozen in the water, and worst of all, you never stop for lunch!'

'It's true,' agreed Omar. 'I've never met people who don't have time for lunch. I think my stomach has gone inwards.'

We had grown close enough to tease one another, and I was sad that in a few days Omar would be leaving. He wouldn't accompany us beyond the borders of Ninewa, he said, because he couldn't offer the same level of knowledge to the south. This was typical of his attitude, even though he could have continued to earn money with us for many more days to come. I asked about his thoughts on the river here.

'It's a living river,' he said, prodding the fish closer to the flames. 'It always changes. I watch it all the time. Different years and days make it seem different. Even the time of day. In the evening when families wash out their rice pans, the colour changes.'

'Are you worried about the pollution we've seen?' Emily asked.

All his life he'd seen the river being polluted, he said. It had always been a problem. He snapped branches, stoked the fire, rotated the fish. Emily asked about his son, and whether he'd bring him to the water.

'Of course,' he said. 'I'll teach him how to swim, but never to fish. I hope his future will be less complicated, and he needs better opportunities. Fishing has no future.'

We ate, gouging out fleshy fish with doughy slabs of circular bread, and rolled directly from the warmth of the fire into down sleeping bags in our row of coloured tents. As darkness fell, we heard hyenas on the opposite banks, their calls merging with the Tigris and the Zab to make one serene soundscape, and I slept the deepest sleep I had anywhere on the river.

Chapter Eighteen

To Be a Hero

Day 26
Safina
River miles: 426

The first sight that greeted us when we emerged from the cocoon of the tents was the long, thin ridge of Makhmour, less than fifteen miles away. The mountain rose alone out of the flat, cultivated, plain, which it puckered at the flanks; a single, winding geological anomaly, like a snake caught beneath a blanket. Makhmour was well reported to be one of ISIS's major remaining strongholds in Iraq. It lay close to a disputed area where the Federal Iraqi army and Peshmerga failed to agree on control. This left a swathe of no man's land, patrolled by neither. In the remote mountains and the caves and valleys within, ISIS lived on, just as they had done on islands and thick reed banks of the Tigris, and out in the desert towards Syria.

On the path that led up into the village, two young men approached us. They'd heard about us, and were environmentalists. When Salman introduced himself, they knew exactly who he was.

'There were wolves here last August,' said one. The

animals had come from Makhmour after a heavy assault by the Iraqi military, fleeing across no man's land towards the Tigris. He showed us a video of a bedraggled wolf swimming across the river.

'Four were killed by villagers, but three survived. We tried to educate people here about looking after them, and not being afraid.'

The two men were in their mid-twenties, well dressed with slick, coiffed hairstyles and prim beards. I asked how they had come to be activists here.

'The internet,' said the first again. 'I looked online, and then later I travelled to Turkey, and I learned we're doing it all wrong here.' The disputed area was a haven for wildlife. They'd seen gazelles and flamingos and more. 'It's a paradise, but people have to stop hunting them.'

As with so many settlements on the Tigris, Safina became a front line during the time of ISIS. Villagers found themselves stuck between Kurdish forces, the Federal army and jihadists coming from the south and west. One activist had a tattoo with four names, each a brother that he'd lost in the war. Just opposite, on a small rise beside the confluence, the army killed six ISIS fighters just three months ago. Unlike other places we'd seen, however, Safina showed few external signs of the trauma. The six hundred homes were rebuilt, and there were paved roads. All four schools were operational again, as was the hospital. Much of this was because it was the village of Sheikh Aziz Sinjar Ezlam Al-Jibouri, one of the most influential leaders of the Jibouri tribe. He had also heard we were passing through and invited us for breakfast.

In Iraq, tribes are organised patriarchally and hierarchically, and every individual in the country is a member of one. Each tribe, or *'ashira*, claims a shared patrilineal ancestry, and that common bloodline creates a bond of solidarity with other members. There are around 150 tribes in Iraq, and their structure generally comprises extended families at the most basic level, which are organised as houses, or sub-clans, and then categorised more broadly into clans. Each clan has a sheikh to lead them, and related clans are led by a *general sheikh*. Identity within tribes is historically strong, and often much more binding than allegiance to any government or state structure. Within a tribe, members will help one another, financially, politically or otherwise, and they may also be expected to help in disputes. These days revenge killings, the most extreme example of tribal obligation, are rare, but do still occur.

Sheikh Aziz was a *general sheikh*, and his home overlooking the Tigris a fortress surrounded by a high perimeter fence. On the roof was a light machine gun, pointing to Makhmour, and by the gate lay a dog the size of a small horse which I was glad took no notice of us. The centre of the compound was the *diwan*, or guest room. It was long and ornately decorated, with Quranic inscriptions above the door. Inside, a patterned carpet unfurled from the entrance, bordered by plush sofas all the way to the far wall. Squat stools, which would be festooned with tea and fruit for visitors, stood like waymarkers on the journey to the Sheikh. He was sitting at the head of the room, on a white leather chair, in white silks with gold trim and dark Ray-Bans. Beside him was an enormous woven wall-hanging of his own face, also in white silks and Ray-Bans. The textiled Sheikh's smile was slightly bigger, and his wrinkles fewer, but its presence gave an impression of the level of power in the room.

We sat under the beaming tapestry beside the real man, and he shook my hand gently. I had been given the seat directly to his right, which was an honour, and as was the custom I tried to decline it and sit further down the hall. Eventually, as a guest and a foreigner, I was placed back at the front, with Emily and Hana on the Sheikh's left, and Claudio and Salman beside me. Omar, who was from the Sheikh's tribe, could not be convinced to come closer, and held unbendingly to his humility, settling uneasily halfway to the exit.

The Sheikh was in his sixties and smelled divine. I did not, and was torn between edging away so he didn't catch my odour and wanting to be close so I could inhale more of his. Intoxicated, I asked, 'What does it take to be a good Sheikh?'

'There are three things,' he said, not missing a beat, and addressing his answer to the room as if it were a press conference. 'The first is that you should be a hero.' He paused to let this sink in. 'The second is that if anyone asks for anything, you should give it to them. The third is that you should understand everything about justice.'

He reckoned there were around six million Jibouris in Iraq, of which roughly eight hundred thousand were directly under his leadership. He relaxed into his seat and spoke again to the room, occasionally flashing a glance at me, but I could not see his eyes; only my own drawn face staring back into his expensive shades.

'Every day I meet with people here so that I can solve their problems,' he continued. 'There's a saying in Arabic that the Jibouri are the salt of the earth, and people come from all over to us to find a solution between themselves or between tribes.'

He built this *diwan* before his own home, he said, and spent more money on it than anything else. The fire outside, where tea and coffee were brewed, should always burn, to show symbolically that his hospitality cannot be extinguished. His had been smouldering for fifteen years, even through war. 'Every day I slaughter at least one lamb. I already have killed one for your breakfast.'

If he was to be believed, he had the respect and admiration of all the surrounding population, with the possible exception of the sheep. He had also worked with the Americans for over a decade and a half. He was an important ally for the US army against Al-Qaeda and led counter-insurgency campaigns in return for military, financial and political support. Many like him hoped their tribal forces would then be integrated into the Iraqi security apparatus, but under the government of Al-Maliki they were overlooked. Often Shi'a leaders were instead posted to Sunni areas to head up security.

When ISIS seized control of territory in Iraq, Sheikh Aziz mobilised his men again. Most Sunni tribal leaders were mindful of the lack of trust the government had previously shown in them and were wary of collaborating with state security forces. In Ninewa, many joined a US-backed programme to train and equip Sunni tribes, which became known as the *Hashd Al-Asha'iri*, the tribal *Hashd*. But Sheikh Aziz instead mobilised alongside the *Hashd al-Shaabi*, who were predominately Shi'a, and formed based on the *fatwa* issued by Grand Ayatollah Ali Al-Sistani. He felt this was the more effective use of his soldiers.

'It was a war of existence, and of ideology,' he said. 'Sunni, Shi'a, we were all together. I had no problem fighting directly under the control of Al-Muhandis.'*

Sheikh Aziz had since withdrawn from the *Hashd al-Shaabi* and spoke of some of the fallout of the war. Their village valued education highly and was known for producing pre-eminent professors, doctors and scientists. 'We now have two hundred men serving in the military,' he told me, 'and a hundred more in the police. We have three pumping stations,

---

* Abu Mahdi Al-Muhandis was a commander of the forty-odd militia forces of the *Hashd*. On 3 January 2020 he was killed by a US drone strike outside Baghdad airport, with Commander of the Iranian Al-Quds Force, Qassim Soleimani.

reliable electricity, sealed roads. We use the river, for fishing, swimming, everything. This is why Da'esh targeted us specifically.' He lost four sons to ISIS. The Sheikh paused and looked at his tea, cooling on the saucer. He touched the glass but did not drink. He wanted us to know that Safina had sacrificed its own blood for everyone. For Iraqis, for Syrians, for Americans. For us. 'We are all brothers,' he said finally, 'sons of Abraham.'

Two men who'd been waiting in the wings now thrust phones at the Sheikh and he held them carelessly, quickly dispatching the callers. Today's business would wait until we'd eaten. A troupe of teenagers wobbled in under silver platters as big as bicycle wheels. On each was a foundation of circular bread with rice, onions and tender lamb that fell from the bone. We sat on the floor, joined by a clutch of tough-looking men with standard-issue military moustaches and bulges in their belts where a handgun was kept. Some fished these out, worryingly cavalier in their doing so, and dropped them on the rug along with wallets, phones, car keys and sunglasses. All across Iraq I saw these identical little piles form at mealtimes, silent markers of machismo and status.

We ate with our fingers, pulling apart meat and pressing it into balls with the rice. When each man was done, he rose and addressed the Sheikh in the customary manner – *Sufra dai'ma, Sheikh* – and staggered to the washroom to clean his hands. When he returned, refreshed, usually with arms washed to the elbow and face glistening from a full scrub, he'd fall onto a sofa and wait to be served tea. Often a young boy, of whom there were many lingering by the door, would be called upon to fetch the handgun that had been forgotten on the carpet, and the boy would be delighted to handle the weapon.

The Sheikh knew a lot about the history of Safina. It

was a source of great pride. Its name in Arabic meant literally 'large boat'. Although the village started life during the Ottoman era as a station for travellers on the river 150 years ago, it gained this current name in 1935. A *safina* made journeys back and forth across the Tigris and serviced the whole area. Later a floating bridge was built upstream of the confluence, but in recent decades it had been washed away and not replaced. Every place name here had a story, said the Sheikh, and he led us out to the fence at the edge of his home. There were two villages across the river abutting the Great Zab. To the north was Jayif, meaning 'dried', because there was once an Assyrian battle there, and afterwards the bodies of the dead were left out under the sun until they shrivelled. That seemed in keeping with what I'd come to expect of the Assyrians. To the south was Makhalat, which was less gory and referred simply to the mixing of the waters.

'The Tigris is the centre of all life,' said Sheikh Aziz. He wanted to see more action to protect it and suggested a campaign. I looked at Salman, and I could see he was making a mental note to remember that statement. But the Sheikh was tired now, and we retired to the *diwan* to sit in bloated silence. As we walked back, he clasped my hand in his.

'The future is bright here. Don't let what you've seen so far tell you otherwise. We will continue to fight Da'esh, and all others, until the end. This is what Prophet Muhammad asks of us.' Iraq had oil, and minerals, and the country was beautiful and ancient, he said. 'We have a lot of friends. Iraq will rise again.'

Before the food had arrived in the *diwan*, Emily and Hana had slipped out to find the women. They were taken through a simple doorway close to the other side of the compound, and there found a different scene entirely. Some buildings were still damaged from mortar shells. The ground was scarred and there was evidence that once there had been a room here, but it must have been obliterated by a rocket.

Eight or nine women, including the two wives of the Sheikh, sat together on a large rug outside, shaded by a corrugated iron sheet, and surrounded by the blood of the lamb slaughtered for breakfast. They wore black abayas, though gold ankle chains and a faint jangle from their wrists suggested much grander attire underneath. Flies swarmed around them and were swatted away from the two large vats of rice and boiled meat. One woman, the wife of another important man from the tribe, dished out the food that was eventually to be served to the men. She smelled of amber perfume, said Emily, and had an off-centre dot tattooed on her forehead.

'Tattoos are *haram*,' she said – forbidden – 'but I have three.' She liked them and didn't care for new religious sensibilities that disapproved. A younger relative, shy and retiring, had tattooed eyebrows, and she nodded along.

Another voice added: 'You know, there was a time when women tattooed their thighs, leading up the inside, because it attracted men.' The women giggled, tugging on their hijabs at this risqué conversation. 'It doesn't work now,' someone responded, and they all laughed louder as she threw up her hands.

The food was delivered to the men, and a woman now working at the sink in the open kitchen was pointed out to Emily.

'She's done well. Her husband is famous here. He was known as the ISIS Killer. He would kill them with his bare hands, and sometimes with a knife or a samurai sword.'

Someone else from the group called over, 'yes, she says he's very good with his hands', and they broke into giggles again.

It always surprised me when Emily reported back on these conversations. Most women in this area would never mix socially with men who were not family, and their roles were clearly defined. They lived behind high walls so no one could see in, and they rarely went out. Responsibilities were mostly related to cooking, and freedoms extremely limited. For me, who never got to meet these women, I might have imagined them shy, or nervous, which is how they reacted if I ever saw them. But when Emily and Hana sat with them, alone and free from the stifling desires of their men, they removed their hijabs and relaxed.

Often, said Emily, they smoked, and spoke of husbands and sex lives. She was never shocked, because those were conversations she had in women's spaces the world over. They'd complain about their bodies, or praise the curves of a younger woman, or covet finer clothing. They'd prod Emily for details of her life. In that one small space, they could be open with one another, but they never saw guests. 'You're the first strangers to visit us,' the woman with tattooed eyebrows told Hana.

There were some who didn't care for the convention, of course. A daughter-in-law of the Sheikh was a nurse and came to tell Emily that she'd worked in Mosul during the war. It was hard work, but thrilling, rewarding even when the hospital was bombed and she lived in and worked out of a caravan for two months. Now she was a GP in Safina, and it was boring.

'It's always a tummy ache or a sore ankle here,' she laughed. But she helped other women here to provide a safe place for widows from the war, and the Sheikh offered accommodation and provided expenses. It was a community, she said, and that made up for occasional boredom in her work.

Hana and Emily took turns to shower in the nurse's home,

and afterwards they climbed a spiral staircase to her bedroom to sit on a golden bedframe and inspect an overburdened make-up table. The nurse pointed out each item – brushes to paint on eyebrows (no tattoos for her), foundation for everyday use, red lipstick for weddings. One tub advertised 'Everyday skin whitening'. She applied it every day and avoided the sun. Even here, even now, the fiction of the superior beauty of pale skin had taken hold.

When they returned to us, Emily and Hana looked and smelled delightful. As we left, the *mukhtar* of the village hugged me, held me tight, and repeatedly gave me the thumbs up. 'Britain, they're the best. The Sheikh loves the Americans. But for me, it's the British. There's always sun on the British Empire, isn't that what they say?' I wondered how on earth he knew that phrase, and I couldn't bear to disappoint the man with my own opinions.

Chapter Nineteen

Honeymooners

Days 27–29
Gayara
River miles: 430

The concept of time and distance for all of us had changed from what it had been before we'd left. That's as it should be on a journey. Travelling teaches you to reconfigure the structure of a day and what can be expected, and miles are only relevant in relation to your mode of transportation. We had moved erratically. Usually it was slow on the river, and we matched the idiosyncrasies of the Tigris itself with meanders and complications. Occasionally we even gathered our mood from its waters, its turmoil or leaden heaviness, and once in a while its quickening, sinuous excitement. We had been following this river for almost a month, and each of us felt the toll in different ways.

More than the rigours of the travel itself, which we were used to, it was the endless negotiations with the military that was most exhausting. I guessed it took on average three to four hours out of each day for us. We were sleeping just four hours most nights and increasingly waking very early to avoid the heat. We spent nights talking until late with hosts. Claudio carved out a couple of hours each evening to back up his video footage, and Emily did the same as often as she could with her photographs. That was about as close as anyone came to down time.

Emily and I had almost no time together alone, unless we were sleeping in tents. On those nights, out of sight and out of earshot, we could have hushed conversations about the day. Both of us worried about the strain the expedition would place on a relationship, but then most of our relationship had been centred around the expedition, and we knew little else. We often said to each other that it was likely no one would be able to make a journey like this again. That surely made it worth the effort. But there were also times, Emily said, that it felt like the rail-replacement bus service of expeditions, by which I think she meant it was just a slog.

I appreciated Salman when we were tired, because he always found a reason to smile. He had charm and charisma in spades and made friends everywhere we went. He travelled under the teknonym of Abu Dijla – Father of the Tigris – and almost anyone we met found his smile magnetic, and his halting, cackling laugh infectious. The direct dealings with *mukhabarat*, army and police often fell to us, as the two men of the group responsible for organising the journey. At times Emily would be involved, too, and her presence was always helpful but, if she realised that the men we were speaking to would not take her seriously, for her own wellbeing and theirs she stepped back.

When Salman and I spoke with these authority figures, whose personalities varied wildly, he did most of the work. Even when I was talking, he would convert the sentiment into his own words in translation, and couch requests in a way that was most likely to work. 'I think of it as adding a little spice,' he'd say. Occasionally, he'd also take heat, as he had done with the *Hashd* after Mosul Dam, or at checkpoints manned by petty authoritarians. He handled these calmly, but sometimes I'd see the grin fall just a little, or I'd catch him leaning on a wall drawing his palm down over his face. He hadn't been sleeping either. Not since Mosul, where he'd shared a room with Claudio, who woke to hear Salman shouting in his nightmares. Now they occurred almost every night and were getting worse. For a while, he brushed it off, but these terrors were becoming a constant companion.

All he would say was that they were related to what happened to him in 2019. In December that year, Salman was abducted by security forces during protests in Baghdad. He would not speak more about it, and we only knew that it was bad. He was held for days. Others like him, young activists, had been killed for their role in the movement at Tahrir Square. He was released, but not unharmed, and after that he left Baghdad with his wife and young daughter. Now they lived in exile in the north. He had returned to Baghdad only once since then. This trip would be the second visit. Our constant interactions with security forces were traumatic and denied his mind the ability to dispel memories of the detention. I knew that in time he might open up more, but for now he continued to work each day, troubled both by lack of sleep and a fear of what would happen when it came.

'Ah, here come the honeymooners!' Lieutenant Zaid was tall and handsome, with a smile much too broad to be genuine. A narrow moustache was forced into a smooth, straight line as he beamed at us. 'Welcome, welcome!' He wore a pristine, pleated uniform. His colleague sported a balaclava, wraparound sunglasses, and carried an assault rifle. They were an odd pair, and insisted on joining us by boat for the next section of river, for our safety. Somehow, Zaid had got the idea that Emily and I were a Norwegian couple on honeymoon, and *Humat Dijlah* was an Oslo-based media organisation filming the once-in-a-lifetime trip. There seemed too big a gulf in understanding to even correct him, so I trusted he'd figure out the truth along the way.

The river ran fast now, emboldened by the surge of the Zab, and when it narrowed to make a bend, the water corrugated and wrinkled the reflected sky. Zaid took pictures of us, and his sidekick adjusted his balaclava. It was getting hot, even when we gunned the engines and pushed into the wind. Every once in a while, we killed the engines and floated. Omar and Ridvan lit cigarettes, and Salman leaned back and closed his eyes. 'I call this airplane mode,' he said, and cast his worries over the side. Daham, we learned, was due to be married in two months. Ridvan slapped his forehead in mock dismay. 'That poor, poor woman,' he said. Daham protested. 'She won't know how lucky she is. Most men can't sing and dance. That's what a woman really wants!' He tried once more to convince us to take him back to England. 'Listen,' he impelled Emily, 'I won't be any trouble. Just take me and put me in the British Museum. I'll sit still!'

We passed just one other boat. In it, a young man reached a long rod into the water, with a wire connected to a generator in the boat. Electric pulse fishing, which sent out a shock into the water and either immobilised or killed fish, so they floated to the surface, was damaging to the ecosystem of the river, and illegal. Salman looked at Zaid, who grinned back. 'It's no problem!' Zaid said, then whispered to him and jerked his head towards us, 'let's not ruin their trip.' It was the first time we'd seen this type of fishing, and it would not be the last. When I asked Omar he said simply, 'idiots'. They had no skill, he added, and no patience, and if he saw them when he was on his own he'd tell them exactly what he thought of them.

We slept a night on a farm outside the town of Gayara. The soldiers lay on the ground outside our tents, hugging their weapons. Under a scarlet sky, Emily spotted an elderly man rowing his boat through the reeds with two grandchildren. She signalled to me and Hana. The low hills on the far side of the river burned orange, and the man made room for us in his boat. 'I love the water,' he said. 'I think that when I die I only wish for it to happen either in the water or while praying.'

'Maybe both,' smiled Hana.

The old man moved his arms like legs on a bicycle, rowing with short quick circles, and for twenty minutes we had the total silence of a river free from engines. Cream slender-billed gulls and black-headed terns glided by, and kingfishers hovered over deeper water. I heard frogs for the first time, and the gentle splash of jumping fish, and in the darkening light they left ripples radiating out across the shallow channel. Soon stars confettied the sky, and the Tigris became just a sound.

Omar and Ridvan left us in the morning. I took a last walk along the riverbank with Omar, and at one point he pulled me by the arm. 'Don't trust anyone,' he said, with sudden urgency. He took my notebook and wrote five sets of coordinates. 'Don't go near these places.' He assured me his intelligence was good. Then he became quiet again, what I'd first mistaken for sullenness but was really deep thought, and a truck arrived to take him and the boats. Zaid left, too, wishing us a happy life, and once more it was just the five of us.

In Gayara, another low-rise industrial town on the Tigris still recovering from ISIS, we rested in the home of a friend of

Omar's. There was no shower, but we could wash ourselves and our clothes with a bucket in the backyard in relative privacy. I'd been getting headaches from lack of sleep, and the dust in the air was making us all cough. Ramadan was beginning, and I hoped it would be a chance for a little rest. The next month would be a time of introspection and prayer for Muslims. Food and water were to be avoided from sunrise to sunset, and the mechanisms of the country followed suit. Shops closed during the day, mosques became busier, and each evening large groups gathered to break fast together. At 3.30 on the first morning we woke before dawn to eat bread, cheese, yoghurt, tahini and tea. We fasted that first day out of respect and spent the next month of travel looking for hidden corners to wolf down snacks.

As far back as the time of the itinerant fourteenth-century explorer Ibn Battuta, who in 1326 had followed the Tigris on land from Baghdad to Mosul, Gayara was known for its bitumen deposits, and Ibn Battuta remarked on how it was exported far and wide. Now it is home to the Gayara oilfield, which holds about eight hundred million barrels of reserves across ninety wells. In 2016, ISIS, who had controlled the area including the oilfield, set fire to eighteen wells as they retreated. The thick smoke, which travelled for over ten miles, limited the possibility of coalition airstrikes and hampered ground forces. It took nine months to put out the fires and cost the life of at least one firefighter. Oil from some wells contaminated parts of the town, too, seeping into streets and pooling on the ground.

I remembered seeing the wall of smoke from Erbil in 2016, some fifty miles away. Our host, Bashir, said that every morning during that time they would wake with black soot covering their faces. At its worst, the family couldn't distinguish day from night. The air was thick, and there was no escape. They called it the 'Da'esh winter'. They had to sweep the house three times a day. Bashir showed us blackening and corrosion on the pipes and air-conditioning units outside. He knew of fifteen people in this part of Gayara alone who received cancer diagnoses since the burning of the oil wells. Although impossible to prove, he saw this as an obvious result of the pollution. 'We live on a gold mine,' he said: 'some get rich, the rest of us get sick.'

The land suffered, too, as oil spilled out around the burning wells. Much seeped into arable land and encroached on the Tigris. Even today rain still washes oil into the river, further polluting the town's drinking source. The attack on the oilfields was part of a 'scorched earth' policy. ISIS also set fire to orchards, forests and crop fields, stole agricultural equipment that could be repurposed to make bombs, and wired homes and farms with explosives. A little farther north, close to Safina village, jihadists set light to the Mishraq sulphur complex, which has one of the largest deposits in the world. Toxic white clouds burned for almost a week, reaching as far as Baghdad, and a stream of sulphuric acid from a processing plant flowed eventually into the Tigris. These attacks caused billions of dollars of financial losses in revenue, alongside damage to infrastructure. The full extent of the environmental and public health impact in northern Iraq is still unknown.

Opposite Bashir's house lived an artist. His studio was bounded on one side by a mosque, and a herd of cattle slept on the central reservation of the road between us. 'There were lots of us before ISIS,' he said when we scuttled across to visit. 'Lots of people with hobbies.' He did a bit of everything but specialised in sculpture. He loved the human form, he said, and inside his studio, alongside a wall splattered with paint, was the wire frame of a human hand waiting to be cast.

Most of his work was destroyed, but he had hidden a few pieces in the ground outside town. ISIS trashed his shop and broke his tattoo gun, which had been a major source of income. Now he made rifle butts and tombstones. The irony of the artist reduced to carving comfortable shoulder stocks for deadly weapons, and headstones for those so often killed by them, was not lost on him. Before we left, he sketched a fishing boat in a river under a bright sun, with a group of cheerful people inside. It was very good. 'Remember the artists of Iraq,' he said.

We had a visit from another member of the *mukhabarat* who brought bad news. There had been an incident, he said, an exchange of fire, and we could not use the river for the next twenty-five miles. I called Omar, who said there was no other option. The coordinates he had given me were also in this area, and there were other spots with high ground overlooking the river, where we'd be exposed. We accepted the decision and hoped for better news further on.

On our last afternoon in Gayara, Claudio disappeared. It took us a couple of hours to notice his absence and, just as we were beginning to worry, he burst back through the gates with Bashir's son, Ali. Ali was a fan of old motorbikes and somehow Claudio and he had come to a wordless understanding of this mutual passion, then careered into the hills on an ancient Russian motorcycle with an off-road sidecar. It was called a Ural and was designed for rugged terrain. Claudio loved it. Not wanting to be left out, we asked Ali to go again, this time with us.

Back in golden light once more, he gunned the engine down a short stretch of the main road, across the old railway line that once carried trains from Baghdad to Mosul, and out over rough undulations by the riverbank. I rode pillion, and Emily and Hana squeezed into the sidecar. Worn leather pilfered from an office chair did little to afford them comfort, and they squealed alternately with joy and pain at the freedom of our ride. Ali revved at high speed through small streams, soaking us from head to toe, and up steep banks until I was sure we'd topple backwards. At the river's edge, where butterflies and dragonflies rose from the water like mist, he stopped, and we all sat there a while. Whatever happened during the day, said Ali, he'd come here and ride until he felt good. As long as there were no fences in the way, nothing could stop him. This was his escape, he said.

Chapter Twenty

Grandmother's Wrinkles

Days 30–32
Shirqat | Assur
River miles: 457

Members of the Iraqi army would accompany us now all the way through Salahaddin Governorate. A man from the *mukhabarat* escorted us from Gayara to Shirqat. He wore a pork pie hat and a polo shirt and had the usual bulge in his belt. His deputy had a beige suit, skinny necktie and wore a pencil moustache and trilby, like a fifties gangster. It was only after a while that I realised they were trying to look normal.

When we had to move by road, our team travelled in a 1980s Kia minibus, because it was cheap and we liked how quirky it was. Inside, ornate tissue boxes were stuck upside down to the roof, in case of emergencies requiring a Kleenex. Behind the driver's head was a framed picture of the bus itself, parked proudly on a busy street in Mosul. It didn't go faster than 40mph, and had no working air conditioning, but these failings we could forgive. The military Humvees topped and tailed us in a very peculiar convoy. Occasionally, one of their vehicles would pull up alongside us and a soldier would lean out of a window to hand us fresh fruit or dates.

This happened on blind bends, fast descents and narrow roads, and I wondered if the biggest threat to our safety here wasn't overenthusiastic military.

It became great fun to have the army around. There were twelve men under the control of a captain called Saif. Most were from Salahaddin, along with a few southerners, and they had been to war together. They saw their time with us as straightforward and before long their guard was down. In the evenings, the men smoked shisha and made video calls back to their families. They were eccentric, and only a couple of them looked remotely fit. A few were distinctly out of shape and one, Hamoud, had a belly so big that he'd often stroke it, and say that the enemy always knew he was coming because they'd see his stomach appearing over the horizon.

What Hamoud lacked in athleticism he made up for with his voice, and throughout the nights he sang a series of classic Iraqi ballads. In between, he'd tell jokes about people from the places we passed through. 'I knew a man from Gayara once,' he said. 'It was 2004, and he decided he needed to make Hajj, so he went to Damascus to get the flight. When he got there, the plane to Mecca was cancelled. So he went to the coast, to the nightclubs in Latakia. In one of the clubs he met another Iraqi, who was shocked to see him. Why are you here, he asked the Iraqi? I'm waiting for the flight to Mecca, the man told him. I'll get rid of all my sins there anyway, so I'm just stocking up!' The soldiers doubled over at this.

On the west bank of the Tigris, less than a mile from where Shirqat collapsed across both sides of the river, a wedge-shaped rocky outcrop pressed the river into a sharp southward bend. The journey there by Kia minibus took less than an hour. Bursts of bottle-green cultivation clung close to the water, and beyond was an ocean of small undulations, barren and beige. At the base of the crag, we arrived at the perimeter walls of a tired-looking building with barred windows and an armed guard. This was the office of Salem Abdullah, archaeological director of the ancient city of Assur, which once thrived on this rock as the first capital of the Assyrian Empire.

This was an area rich in history. Thirty miles east was the UNESCO World Heritage Site of Hatra, a two-thousand-year-old caravan city from the independent empire of the Parthians, and an important juncture on the Silk Road. ISIS had held both Hatra and Nimrud further north. During that time, they released videos from the latter of jihadists blowing up the three-thousand-year-old Temple of Nabu, the Assyrian god of wisdom. Other clips on social media show pneumatic drills and sledgehammers being used on Assyrian friezes, and in Hatra the same methods were used there. UNESCO declared this 'cultural cleansing' and UN Secretary-General Ban Ki-moon called it a war crime. Two Humvees and twelve soldiers accompanied us to Assur, but once there they were content to wait at the office and let us wander freely.

'My relationship with this place is above a job,' Salem told me as he welcomed us to the office, lines creasing the corners of his eyes as he smiled. He was born in a village close to Shirqat, and from his house he could see the site. His father worked at Assur, where he died of a heart attack among the ruins, and since 2001 Salem had worked there, too. 'It is like family,' he said, speaking in lilting, classical Arabic that even had Salman on the back foot as he translated. 'I think of this place like my grandmother. Come, and I'll introduce you.'

We left the office under a blazing summer sun, stepping over the remains of mud-brick walls and slowly climbing as we crossed the city. A hot wind kicked up ancient soil. The site measured only a square mile, but walking it made it feel larger. Salem avoided the dirt road that cut through the centre, instead picking a path straight over the ruins. He wore a button-down shirt and grey polyester suit, but scrambled over low walls and trenches with ease. Finally, we stopped at the edge of an escarpment, mopping our brows. The Tigris tumbled by, eighty-five feet below. To our right a crumbling ziggurat rose from the lip of the cliff. 'This area was the temple of Assur,' Salem said, shielding his eyes from swirling dust. 'The most important place in the whole city.'

The Assyrian Empire grew out of the founding of the city-state of Assur in the third millennium BC. Assur, the empire's first capital, was believed to be the physical mani-festation of the deity for whom the city was named, and the temple his eternal residence. But it was also a wealthy hub for regional trade, positioned along a main caravan route, and it formed a lucrative trading relationship with Anatolia in what is now Turkey. Much of what we know about the city's early flourishing comes from a remarkable collection of over twenty-three thousand clay Assyrian tablets discovered at the Turkish site of Karum Kanesh, six hundred miles away.

Traders to Karum Kanesh would have followed the Tigris north to Nineveh, then traversed the foothills of the Taurus Mountains. Donkey caravans carried mostly tex-tiles, made locally and imported from Babylon to the south, and tin, which had come through Iran from central Asia. Gold and silver from the sales in Anatolia would return to Assur. Elsewhere, traders also made purchases of wine. The Assyriologist Karen Radner writes of a contemporary of King Ashurbanipal, who sent four caravan leaders three times a year to Sinjar mountain. In antiquity, this area was famed for its wine and once the goods were sold, wine was bought with the proceeds. Radner says: 'The wine was filled into sheep or goat skins. As these skins were traditionally used for

buoyancy on *keleks*, this created a happy dual purpose. The
wineskins were bound with logs to create rafts for the return
journey to Assur on the Tigris. This was good for the wine,
as the river kept it cool and prevented it from spoiling.' Back
in Assur, the wine was moved to cellars for the well-to-do in
the city, and the logs from the rafts repurposed into timber
for construction.

As well as the caravan route, there was another way called
the King's Road. It ran the length and breadth of the empire
and was likely the innovation of Shalmaneser III. Each
Assyrian region maintained roadhouses along it to provide
overnight accommodation and resupply to messengers and
envoys of the king. The closest equivalent is the caravanserai
of the Silk Roads. On the King's Road, however, only travellers bearing the royal seal were allowed access. This system
for rapid, long-distance communication between the king
and his administrators became adapted and used by subsequent empires and became one foundation of Assyrian power.

Despite this, Assur was never a big city. The natural
defence provided by the Tigris and another arm of the river
to the north in antiquity were supplemented with two layers
of fortified walls in the south and west, enclosing a modest
but easily protected nucleus. The residents spoke a Semitic
language called Assyrian, closely related to Babylonian,
and even when King Ashurnasirpal II moved the capital to
Nimrud in the ninth century, Assur remained divine and
prosperous. That was true right up until the sack of the city
in 614 BC, by the Medes, which ultimately brought an end to
the dominance of the Neo-Assyrian Empire.

As I looked with Salem across the site, the ancient city
took shape before my eyes. Ruins coalesced like a blueprint
into housing districts, temple walls and occasional monumental buildings. Most dramatic was the ziggurat, which

is some eighty-five feet tall and once stood at least twice as high. More than four thousand years old, it was part of a temple complex dedicated to the god Assur. Its six million mud-bricks were once covered with sheets of iron and lead and inlaid with crystals. Now the great mound looked as if it were melting, with dried mud settled like candle wax around the base. On the side, a large opening led to the interior. Salem was quick to point out that this was not part of the design. Hormuzd Rassam had bored the entrance, he said, to look for artefacts. Layard and Rassam were among the first to dig here, and turned up a life-size carving of Shalmaneser III, surrounded by cuneiform. It was the first ever Assyrian statue to be found.

'Only a fraction of all this has ever been excavated,' Salem said, looking back across the city. His estimate was that 85–90 per cent of the site remained unexplored, and he believed there was much still to learn from Assur about the ancient Middle East. 'If more work is done here, it will change history. There were one hundred and seventeen Assyrian kings. When these kings died, they were buried here,' he said. But to date only three royal graves have been identified. 'Where are the rest?' He paused. 'They're here, under our feet.'

With the exception of Layard and Rassam's cursory explorations, Assur was mostly overlooked by British and French archaeologists in favour of Nineveh and Nimrud. It wasn't until the turn of the twentieth century, when a German expedition led by Walter Andrae established the city boundaries by cutting a series of trenches across the site, that more of the city's structure and life became clear. Andrae and his team were at the forefront of a more scientific approach, and although progress was slow, they caused significantly less damage than their predecessors.

The archaeologists recovered thousands of cylinder seals and baked clay tablets, some carved with cuneiform inscriptions written in the second millennium BC, which detailed religious rituals, business transactions and other subjects. Most of these artefacts were shipped to Berlin, where they are still on view at the Pergamon Museum, or to Istanbul, then the seat of the Ottoman Sultan. In recent decades, excavations have been intermittent. 'For Iraqis, it's expensive,' Salem said. 'The government can't afford it.' The last major international-led excavation concluded in 2002. 'And here we get to the major issue,' he sighed. 'There are many positives to this place, but always there will be challenges in greater numbers.'

It's proved nearly impossible to secure the site. A mesh fence runs along to the road, but many sections have been flattened or removed altogether. And while a visitor technically requires a ticket, without staff to enforce the rule that system hasn't worked for thirty years. Instead, residents of Shirqat treat Assur like a local park, wandering in for picnics. 'In spring you can't see the ground,' said Salem.

During the ISIS occupation, Salem and his staff fled, burying their archives in the garden of a friend. Damage to homes and infrastructure in the town was enormous, and many are still deeply traumatised. Salem believed Assur was a source of great pride for most Shirqatis, and that the healing of the site and the population could happen together. 'The reason they come here is because they love it,' he said. But they needed educating on how to interact with the site.

There was looting, too. Every time it rained, topsoil was washed away and artefacts – potsherds and even cuneiform tablets and statuettes – emerged from the ground. In 2018, heavy winter rains caused the Tigris to flood and in the aftermath 180 artefacts were revealed, lying plainly in sight.

Shirqatis gathered them and took them to Salem, who made sure they reached the antiquities department in Tikrit. But although Salem believed most Shirqatis respect the site too much to steal, it wouldn't be difficult to pick up a few things and traffic them on the black market.

ISIS ran a lucrative trafficking industry, with looted artefacts, jewellery and more making their way from Iraq and Syria through Turkey or Lebanon to Europe to be sold on to Western buyers. Salem didn't want to say more on this but later, in Shirqat, a man who didn't want to be named told us the trafficking continues, and the same people are still involved. He agreed with Salem that most Shirqat residents understood that their own heritage was tied to the ancient civilisations, and had a sense of the need for protection, but even with a few smugglers, much was being lost.

Alongside Assur, Salem's role also included managing all the archaeological sites in the Shirqat area, which numbered 274. 'In total, we have four archaeologists and nine security guards,' he told me. 'Every guard has to look after ten to twelve sites, and the other locations have nothing.' He shrugged and we walked again, now past the headquarters where Walter Andrae had been based. Broken glass and rubble were strewn on the ground, and what remained of the walls were streaked with graffiti. Inside, a single window looked out on the Tigris. When British archaeologist Max Mallowan visited in the 1950s, his wife, Agatha Christie, accompanied him. She saw this window and was inspired to write, said Salem. 'If it was up to us, we'd build a monument to her. We love her. She believed in Iraq.'

To the west, the three broad arches of the Tabira Gate glowed like bronze in the amber light of early evening. The structure is the best-preserved monument at the site, and probably dates to the fourteenth century BC. One theory is that it was a processional route for people and gods on their way to the ziggurat and temples. Another suggests it was the gate of war, only to be used when the Assyrian army marched out of the city to or from battle. In May 2015, ISIS militants blew a vast hole in the structure. The damage was estimated at 70 per cent, but in 2020, after the area's liberation from ISIS, a joint project between the American University of Iraq, Sulaimani and the International Alliance for the Protection of Heritage in Conflict Areas, known as the ALIPH Foundation, carried out emergency reconstruction work. By the time I arrived, the contemporary sun-dried mud-bricks had bedded in nicely.

On our way back to his office, Salem stopped to outline his biggest concern for the site. Twenty-five miles south, the government is planning to construct a new dam at Makhoul. Amid growing fears of dam projects in neighbouring Turkey and Iran exacerbating water scarcity in Iraq, Makhoul purports to offer a strategic solution. The Ministry of Water Resources expects a storage capacity of 105 billion cubic feet of water, which will primarily irrigate the surrounding agricultural areas in dry seasons. Originally the dam was proposed in 2002, and when UNESCO named Assur as a World Heritage Site in danger the following year, the agency cautioned that the reservoir could flood scores of archaeological sites around Shirqat. The project was halted by the fall of Saddam Hussein, and the current, revived plan is plagued by the same problems as before.

The Head of Archaeology at the University of Tikrit, Khalil Khalaf Al-Jbory, had been hoping to join us but was pulled away by work. We spoke instead by phone from Assur. The geological foundation of the dam site is soluble, he said, like at Mosul, and sulphur seepage will contaminate the water. He also pointed to what he called a 'social disaster', with tens of thousands of people facing displacement. Khalil believed that over two hundred archaeological sites were at risk of flooding and infiltration. He had done the work to map these himself, marking their elevation against that of the proposed dam.

Assyrian sites, constructed primarily of mud, stand to be lost for ever. In 2002, there were proposals to build a retaining wall to protect part of Assur, but now so little information is available that even those closest to the sites are left in the dark. 'The government is not listening to anyone,' Khalil told me. 'Not to the academics, or geologists, or anyone. It's very dangerous, and very risky.' Salem shared similar fears but, perhaps because his employer was the government, he was more considered with his language. He had not lost hope but agreed that Assur's future was dire unless something could be done to alter the plans. 'When I say this is my grandmother, I mean that I also see her wrinkles,' he told me. 'She needs help now.'

Chapter Twenty-One

Submerged

Days 33–35
Subayh | Makhoul Dam site | Baiji | Al Alam
River miles: 528

In Shirqat we pitched our tents in a family garden while the soldiers piled into a guest room. The garden overlooked the Tigris and in the morning I swam with a soldier called Tarik. I felt the pull of the current when I swam out from the bank, and in a moment of clarity I realised what it was to be swimming here, in the waters of the Tigris, close to ancient Assur. There it felt hopeful that all we'd seen coming undone in the country could somehow be stitched back together. We had met so many good people, and surely something as powerful as this river, something with the potential to be consistent and resilient, was part of the solution. There is nothing like cold water immersion to bring about idealism in the early morning.

We found boats to rent close to Assur. The mountains on the right bank were squat and uneven, the ridgeline gnawed like a fingernail. We had two soldiers in our boat, and the rest travelled separately by road. The young men with us watched the slopes and spurs carefully. Our position was vulnerable, but I was happy to see we'd got two of the more serious soldiers.

At a village called Subayh, the local sheikh came to sit with us under the shade of his son's porch. The house was pale orange, and fruit trees grew by the road. The sheikh had not heard anything official about the dam. 'We get our news from Facebook,' he said. 'I ask my son to look each day and tell me what's happening.' There had been no delegation and no survey. He had asked friends in other villages, and they'd heard somewhere that this area would be flooded in the first phase, within four to five months of the dam being impounded.

There were 250 families in Subayh in 2014, he said, and they'd already lost a lot to ISIS. Now they feared their homes would go, too. 'What can we do?' he asked. 'My father and grandfather are from here. This is my home. Can I move it brick by brick? What about my mum? It would kill her to have to move.' I thought how often I had heard this, how often someone found themselves unable to continue the lifestyle of their forefathers. He said that at least if they received a payment they could go somewhere else, but he didn't believe this would happen. 'The worst for me would be to end up in an apartment,' he said. 'I grew up here, planting crops, swimming in the river. What would I do in an apartment?'

We visited several villages due to be flooded and none had been contacted by the government. One man said he hadn't taken the dam idea seriously, but if there were foreigners asking questions, maybe he should. In another, a trainee lawyer in his early twenties took us to a lookout by a bend in the river. Behind us, palm trees gave enough shade for children to chase a chicken with a football. Reeds fringed the edges of the water, and there was a single fishing boat. The lawyer wanted to have a family here one day. It was paradise, he said, but it too would be gone. His pain was at its most palpable in his shy speechlessness, when explanations no longer came. He pulled at his *dishdasha*, and a hot wind blew in the smell of meat cooking somewhere close by.

Close to these villages we reached another waypoint, where the Lower Zab River joined from the east. It had come from the high Zagros Mountains in Iran, also carving an improbable descent through deep, lateral canyons. For a while it was the border between Iran and Iraq and it was dammed twice in Iraq, with a further Iranian dam under construction. Once it was famed for its prodigious flow, but, like the Tigris, it now reflected the general thankless toil of Iraq's waterways.

We climbed up on a small promontory to watch it struggle in. Here it was the Tigris that was muddy, and the Lower Zab a pale blue thread. In Shirqat we had spoken to a head of the local authority, who estimated the Lower Zab's flow was 90 per cent down on a decade ago. He also said there were seven thousand houses, in twenty-seven villages, that would be under the water if Makhoul Dam proceeded. The natural shape of the confluence would be lost, and where we stood would be a hundred feet under a lake.

To the west began the Makhoul Mountains, bleak and dry, running as a single, narrow spine parallel with the river. Not far downstream, these would form one side of the dam, and lower hills presumably risen and reinforced on the other. At the pinch point, the skeleton of a bridge stretched out towards the eastern side, but in the middle dropped away suddenly so that all that was left were girders like fangs pronging the water. It had been destroyed by Al-Qaeda and there had been no money to rebuild it. Below was a causeway of concrete. Tarik, the soldier who liked to swim, said the causeway would assist in rebuilding the bridge, and then the bridge used to help with dam construction. Underneath, there had already been a foundation built during the previous iteration of the project. The river ran erratically, responding to sub-aquatic disturbances.

On the far side, three excavators pushed sand and gravel into the Tigris. This was ground zero, and beside the road on our side of the bridge, a brand-new cornerstone had been placed, regally overlooking the area. Under a line of Quranic blessing, the names of the prime minister and the minister of water resources, it noted the date according to the Islamic Calendar and that it marked the implementation of the pro-ject. It had been laid four and a half days earlier. We had just missed it. I wondered why the ceremony for the cornerstone, for there surely was one, did not include an invitation to the local communities that we'd met. Salman guessed there was now already a budget and a signed contract.

Tarik went swimming and Salman and I sat on the bridge, watching a shepherd drive his flock across the mess of con-crete and remembered conflict. The sheep navigated it like the mountain and moved smoothly. The air was dusty and dry, and the river churned. Salman said there were so many other options to store water in Iraq. Why not use the marshes in the south, he asked? And why build another unstable dam like Mosul, and create a second hazardous structure that put Iraqi lives at risk?

'All the other countries shout about Da'esh, because it threatens them, but what about these dams? The Ministry of Water Resources has international advisers who are support-ing these projects. Where is the international community in helping us with this?'

Humat Dijlah had requested information on the budget and proposal, but nothing had been released. Salman kicked a couple of stones off the edge and watched them land below.

'This is the most horrible area I've been to in Salahaddin. The effect of this on the biodiversity, water quality, environment, I can't imagine it.' It was the first time I saw his optimism drained and, like the bridge we sat on, for a short time he saw no place to go.

Although it was Ramadan, we never went hungry in the company of the soldiers. They had friends everywhere along the river, and each day arranged someone to host us for lunch. Their exemption from fasting was justified by working, so in the middle of each day we'd be presented with platters of *maqlubeh* – spiced rice and roasted vegetables and chicken, cooked together and then flipped upside down to serve – or *tshreb* – layers of bread soaked in a yellow turmeric or red tomato sauce with cuts of lamb.

The soldiers would discuss the food at length. The captain had recently had a gastric bypass and dropped thirty kilograms, so he couldn't eat as much as he liked. Others would detail the spices involved, and where the best ingredients came from, often arguing until the host came and settled it. The men of the family whose home we were in would serve us, running in and out with trays, and refilling tea and offering cigarettes. Emily and Hana would split their time between the men in the guest room and the women in the kitchen. Hana, who was vegetarian, could prepare herself bespoke dishes, in stark contrast to the pickings she sometimes had to make do with. Emily enjoyed these days greatly, and was energised by the time we spent with the soldiers and their families. Once the meal was finished, and cigarettes smoked, the soldiers would lie down on soft blankets along the sides of the guest room and fall asleep. Some would use the legs or torsos of others as a pillow. We'd nap together for an hour, then move on.

In the evenings, something similar happened for *iftar*. It was understood that we should finish our day's work and travel before sunset so that we'd be ready to eat as soon as the sun set and the call from the mosque was heard. These meals were ludicrously large. Even if we had been denying ourselves all day it would have been excessive, but after a heavy lunch our team struggled to make inroads into the columns of rice and meat.

I hadn't expected to be so well fed on this part of the journey, but I liked the routine and I enjoyed the company of the soldiers. They were fun, and as they grew to know us they shared their aspirations. One wanted to be a teacher, another to run a café. A third was going back to Baghdad to study engineering. In the mornings they'd shave in pocket mirrors, spending the most time trimming moustaches, and they were always respectful of Emily and Hana and their need for a little space from the mass of sticky men. The only downside to their company was that there was usually just one toilet in each home that we stayed in, and a race ensued in the morning to visit first. If you slept in, as I did once, the smell that emanated from the outhouse was nauseating. I felt ill for the rest of that day.

The road to Baiji switchbacked over the Makhoul Mountains and out onto a dry plain. Away from the Tigris, colour drained from the land. Travelling by car was always a reminder that however bad the situation on the river, it was exponentially worse the further inland one went. In Baiji, another battered town on the river, the soldiers left us for a few hours. Once this had been an important junction on the railway track, connecting the Baghdad–Mosul track with an artery that punched through the desert to the west. It was becoming common for us to see these tracks, and there had been consistent talk from the government in Baghdad of reviving the line to Mosul, and even on to Turkey, but nothing had happened.

The previous year, Emily and I had taken the passenger train from Baghdad to Basra, which is now all that remains of a once rich heritage of rail travel in Iraq. A tram system was replaced in 1914 by an Ottoman track running north from Baghdad, and eight years later the invading British constructed a narrow-gauge line to connect the southern port city of Basra. In 1943, the first continuous journey was made from Istanbul to Baghdad. The trip took three days and was the culmination of thirty years of imperial attempts to connect Europe to Iraq via the Turkish capital. The railway was also a major reason for the decline in importance of the Tigris and Euphrates as tools of transport and trade. Much of the line ran along the general route of the Euphrates and, once goods could be loaded onto railway cars, the river was no longer so desirable. Roads followed in time. Today, after decades of conflict, sanctions and occupation, most of the railways in the country lay abandoned, like here in Baiji, lifeless and skeletal, buried under sand, mud or water.

The major industry in Baiji was the oil refinery. It had been the largest in Iraq, and although it was severely damaged in the fight with ISIS, it was now operational again, albeit at a reduced capacity. The air was still tangibly thick and noxious. In the distance we could see the orange flames of the flare stacks, which burn off by-products of extraction. Their hue bled into the sky, leaving a beige smear over the town. The refinery was a major source of concern for Salman and other environmentalists. Refining oil is complex, requiring various phases of cooling, hydro-treatment and desalination. Often wastewater from these processes finds its way back to the river, said Salman. He had been part of water-quality testing teams in the past, and they had found chemicals and heavy metals* in the Tigris, almost certainly derived from oil spills and mismanagement of waste. The agricultural land in the area has been severely degraded by pollutants, too, and there were innumerable reports of illness in the nearby population that they attributed to the impact of the oil refinery.

'The problem gets bigger and bigger, the more you look at it,' said Salman. He said we would see more refineries along the banks of the river as we went south and also mentioned the abandoned toxic waste that littered the country. Much of it was the legacy of decades of war, he said, and in particular the American occupation. 'There are laws, but they're being ignored. Then there's everything from the past.' If there were still sites contaminated with depleted uranium, he said, what were the chances of enforcing better environmental standards in the oil refineries?

The market in Baiji was busy, and vendors bagged up cucumbers and garlic and potatoes and herbs for the crowds stocking their *iftar* ingredients. Emily wandered off to buy some vegetables. She was trying to be relatively inconspicuous, but that was ruined when Tarik arrived, standing on the roof of a military pickup truck wearing a T-shirt that said '*All Guns Matter*'. 'Hi Emily!' he waved, grinning, unaware that everyone was now looking at him, then her. He was bored and wanted to chat. She blushed, her cheeks matching the tomatoes, and scurried back to the rest of the group under a heavy sky.

We returned to the river with a policeman, Nazar, who had under his control two fine white fibreglass police boats. Each had a large machine gun mounted at the bow. Two tired Iraqi flags flew valiantly at the gunwales. Nazar apologised for the guns, strapped on a life vest, and flicked an ignition

---

* A broad term to include all the toxic metals.

switch. This was a real upgrade. Now we had 175hp, and
our soldiers were happy to leave us, probably because of the
machine guns.

Nazar liked the middle of the river, so for a while all
we could see was the gentle motion of reeds on either side.
Fishing was still forbidden for a few more weeks because
of the spawning season, so when we passed a boat with a
generator, I imagined Nazar would want to enforce the law.
But if his willingness was there, his power was not. The two
men in the boat laughed when we stopped, and one immediately called a number on his phone. Within a few moments,
Nazar's phone was ringing, too. These men were with the
watermelons, Hana translated, and should be left alone. Their
militia outranked Federal police in this scenario. In a microcosm it showed the confusion of control in Iraq. The men
grinned smugly and did not wish us well as we left.

In deeper water, a few resourceful villagers had made small,
square walkways leading out from the bank. Underneath
were fish farms filled with squirming carp ballooning in
captivity. Beyond, the fringes of palm trees were visible. Both
the farms and the palms were to become frequent sights, and
a sign that we were entering a new phase of the river. Salman
and I made the familiar observations, too, of the pumping
stations that were a satellite for each village, and the sporadic
effluent pipes.

On one of the long, oval islands, the boats slowed so that
Nazar could point out a slight depression in the reeds. 'Wild
boar,' he said. 'There's a lot of them on these islands. The
Christians upstream in Mosul would eat them, but we just let
them roam. If they annoy anyone, they get shot.' He seemed
happy with this arrangement and smiled at a memory which
wasn't shared.

In the town of Al-Alam, Nazar docked by a rough bank,

and we scrambled up the steep slope to a thick eucalyptus tree. There, waiting for us, was a woman called Um Qusay, who had inadvertently become a symbol of the resistance against ISIS. Emily had made contact a few days prior, and Um Qusay arrived early to wait for us in a long black abaya sprinkled with gold detail. She was ready to tell us her story, she said, and sat down under the tree. Her request was only that we stay in the shade, because she was fasting for Ramadan.

In June 2014, ISIS tore through the city of Tikrit, just south of Al-Alam. In the process, they routed a large Iraqi air force training base called Camp Speicher. The recruits were mostly Shia, all rookies, unarmed. ISIS rounded them up easily and took the prisoners on buses into Tikrit where they were executed and dumped in the river. That was perhaps the most disturbing type of pollution we encountered on the Tigris: the mass graves. The exact number of dead from Speicher is unclear, but as many as 1,700 young lives were lost in the single bloodiest massacre that ISIS carried out. It was also the second deadliest terrorist attack in the world after 9/11.

Those recruits who escaped, some 850, fled north along the Tigris. Trapped on the west side, surrounded by danger, the river lay between them and the possibility of safety. We had heard this before, elsewhere on the river. To so many, the Tigris had been a barrier between life and death, and the kindness of strangers the only possible bridge. For the recruits, they needed the compassion of villagers in Al-Alam on the other side.

Um Qusay sat with us now, looking out across the wide, lethargic Tigris. 'This river was the front line,' she told Emily, nodding slowly and squinting into the sun. Her eyebrows, drawn on in perfect black lines, furrowed ever so slightly.

'The soldiers shouted to us. They knew we would help. So I sent my son in his boat to bring them back. He would bring back five or more at a time, and he kept going.'

The soldiers arrived sporadically on the far bank in small, ragged groups, and when her son delivered them to safety she was the first to greet them. Later, when the jihadists caught up with the fleeing recruits, other residents from Al-Alam provided covering fire from where we now sat to give the boatmen time to make a dash across.

'When any of them came, I would hug them. It was like I could receive my own family again.'

Um Qusay's husband and another son had both been killed at a checkpoint by ISIS early in the conflict. For her, the chance to save the lives of others was a gift, and though it would not bring back her family, it could save others from the pain she knew.

'I welcomed everyone like they were my own husband and son, and invited them inside. I gave them clothes, and let them bathe, and I cooked them food, and I made them feel good.' She was sure she had hosted at least fifty-eight young men. Some stayed for days before moving on.

She took a moment to pause, and draw breath, and take in the air. Normally she'd smoke to feel calm, but it was Ramadan and cigarettes were off-limits. 'Sitting near the river still hurts me a lot,' she said. She reached out for Emily's hand, and they held each other's for a while. Now the corners of her eyes were damp, and when she blinked tears flowed. She dabbed at her cheeks and exhaled.

'I think of my husband and son, fishing here, having a good time here. I used to swim too in private places. Now I can't look at the water. I think of ISIS throwing bodies here instead, and all I see is death.'

Many others in Al-Alam helped the soldiers, too, she told

Emily, but other families were afraid of making that widely known for fear of retaliation from ISIS in the future. They kept a low profile, but Um Qusay had already lost so much.

'When the *Hashd* came and cleared out ISIS, I jumped on the tank. Every day I went with them, and soon I was telling them where to go to find the terrorists. I had so much revenge in my heart.'

Eventually the *Hashd* media heard her story, and she was approached for an interview. Then another, and another. In 2015, she was honoured with the Medal of the State in Iraq and, in 2019, the International Women of Courage Award, awarded by Melania Trump. Still, her heart ached. Still, what was taken from her would not come back. But she dreamed of better times, and she was proud that what happened at Al-Alam could show Iraq was not divided along sectarian lines. Not when it mattered. 'One day,' she said, handing an empty teacup back to her son, 'peace will come to here. And then, then I'll rest.'

Chapter Twenty-Two

People's Palaces

Days 36–38
Tikrit
River miles: 537

Nazar took us in the police boat to Tikrit where large, luxurious homes perched atop steep, crumbling banks. Housing districts rolled like carpet laid over the right side of the river. Some had stairways winding down to the water, where slender jetties propped up private fishing boats. Others had seen their days of glory pass, and crenellated walls were partially collapsed, sinking down the slopes with the flaky soil.

This city had once been a place of great wealth and power. That much was clear. Saddam Hussein, a son of Tikrit, drew a majority of his close Ba'athist cadres and elite Republican Guard troops from his own tribe in the city. Many of the sprawling villas would have been built then, and some destroyed in the insurgencies following the American invasion. More still were flattened during the time of ISIS.

Nazar left us on an island in the Tigris, within walking distance of the first palace in Saddam's presidential complex. During his reign, Saddam infamously built dozens of garishly extravagant palaces around Iraq. I had seen some before, on

top of Gara Mountain in the Kurdistan Region, and overlooking the archaeological site at Babylon. Now, for the first
time, I saw the heart of his autocratic stronghold and, like the
satellite palaces, these also lay in ruins. In Tikrit alone, there
may well be over a hundred mansions, built for Saddam, his
almost impossibly evil sons and various others close to them,
often with swimming pools, lakes, orchards and always –
always – gaudy, brash decor.

After 2003, the palaces came under the control of the
Americans, then the Iraqi army, then ISIS, and now the
Hashd al-Shaabi. We were allowed access, briefly, and hurried
off the island towards the closest of the palaces. Today, any
sense of luxury has been stripped from them, but still they
stand, some tall and beige, others buckled and bowed, surrounded by palm trees and battered concrete walls.

It's said that Saddam Hussein imagined himself like the
ancient Mesopotamian kings. In particular, he identified
with Nebuchadnezzar II, channelling him in his speeches,
and in some portraits his facial dimensions were even
altered to match the impression of the Babylonian. Images
of him riding to battle on a chariot developed the theme. At
Babylon, he spent millions rebuilding the ancient site, using
bricks with his own name inscribed on them, just as his
warrior-king hero had done.

There were elements of that Babylonian admiration
present in the palaces in Tikrit, alongside other symbolic
architectural choices. One was named after a mythical sword
and shaped appropriately; another had a series of disembodied
hands linked around a pillar, apparently to represent the ethnicities and religions of Iraq co-existing. The building next
to it had been levelled by an airstrike, and the *Hashd* soldier
watching us did not appreciate the irony of the Sunni, Shia,
Christian and Yazidi fingers intertwined in such a theatre.

Inside the building of co-existence, the marbled floor of a colossal hallway lay under a blanket of smashed glass and fragments of sandstone. The walls were riddled with bullet holes. Around the sides an arcing staircase, also marble, climbed gracefully to a mezzanine level. A shattered chandelier hung precariously in the centre of a tiled roof like the sword of Damocles. All these palaces had been looted, and I'd met Iraqis in the past who had taken keepsakes for their homes.

The hall had three arterial rooms. In one, threadbare sofas were laid out alongside simple sleeping rolls and floral blankets. An old cabinet was being used as a clothes horse and, facing away from us, staring at a wall scrawled with anti-Saddam graffiti, two *Hashd* sat smoking cigarettes. They lived here on rotation, like any other military base. I'd heard talk of the palace buildings being turned into tourist attractions, or museums, but based on what I saw there was little left to salvage.

The *Hashd* soldier who had been watching us outside came in now and stared hard. He was drunk, Salman thought, or on drugs. His pupils were dilated, and he didn't seem to have full control of his limbs. Sometimes soldiers took stimulants to keep them awake, Salman told me later, because they worked such long shifts. Others were just addicted, or had become so during the war with ISIS when falling asleep could have meant death.

'He called these the People's Palaces,' said the man suddenly, his legs unsteady, a hesitant smile on his lips. 'But Iraqi people couldn't come in. We didn't even have food. So how could it have been for the people?'

He stood back, a little triumphant, rolling on his heels.

'But aside from the politics, there were good things back then,' he said, frowning now. He waved his hand towards himself, as if beckoning in the past. 'It wasn't great, but there were good things. Anyway, there's something very peaceful I feel when I walk around here.'

We often heard reminiscences like this. Although now it seems strange for a Shia militia to think fondly of Saddam's time, it has to be remembered that the sectarian divides were not apparent then as they are now. All that, Salman kept reminding me, came after 2003. Before there was still Sunni and Shia, he said, and other religions, but during Saddam Hussein's rule everyone lived under authoritarianism, and sect offered little help if you stepped out of line.

We walked to the concrete platform where ISIS militants had killed the Speicher recruits. The river coursed past a lower palace, capricious, and I could not look at it. Even the sound, so often tranquil, even that was unbearable. I thought of Um Qusay, living beside the Tigris, haunted by water. The walls were plastered with posters of dead men's faces, often superimposed over the holy shrines of Karbala. At the water's edge a bronze sculpture showed three mourning widows, or mothers, digging in the soil only to find the skulls of their loved ones. Beside it, an enormous poster that showed stills from the videos the jihadists posted of the massacre online. Another *Hashd* soldier said, 'If you think this is bad, try coming back next month. When it gets really hot, you can still smell the bodies.'

When we left, a man in a fishing boat told us about the hundreds of men who had disappeared at the hands of the militias since the liberation from ISIS. 'They were taken right here, by the river,' he said of two friends. The militias had been responsible for an undocumented number of extrajudicial killings and kidnapping, as well as the destruction of property, and we had heard similar stories all throughout the Sunni areas on the river.

I was not sure we could hear any more. Our group morale was low, and only Claudio seemed able to protect himself emotionally from what we'd seen. Emily confided that she desperately needed a break. She was at risk of losing her faith in the goodness of people, she said. For five weeks, we had been on the move. Any single day, like in Tikrit, could be filled with half a dozen stories of trauma and loss. I had always felt a responsibility to listen to these personal accounts, because those who had suffered deserved to be heard, at the very least. But it was overwhelming us, and we were all strung out. There had been little good news from the Tigris recently. The soldiers with us worked only five days at a time and then rotated out to take a three-day break. Captain Saif said it was tiring for them to work such long hours as we did, and sleep so little. Salman's nightmares were worse than ever, and Hana was suffering, too. She had been Salman's primary source of support, and in helping him work through the night terrors, as if by osmosis, she had now absorbed them herself. I could not imagine experiencing someone else's nightmare.

We drove back through the city past the birthplace of Saladin, founder of the twelfth-century Ayyubid dynasty and the man for whom the governorate was named. Close by was a seventh-century Syriac church known as the Green Church. Tikrit had a celebrated Christian heritage, which existed now only in tumbled remains. The Christians survived in the city until the seventeenth century, through successive Islamic dynasties, and even through the Turco-Mongol conqueror's razing of the city in 1393. A small Jewish population was still in the city in the 1950s, but disappeared soon after. Tikrit was so much more than Saddam Hussein. But now it was his name, and that of the Islamic State, that dominated. This had been a pattern throughout: deep history was buried under the recent woes.

We broke fast with friends of Salman's. Samar and Ibrahim lived in a smart house in a pleasant part of town, with high walls and a fragrant rose garden. The tiled pathway was spotless, brushed clean of the dust that was perpetually in the air, and at the threshold the couple waited for us. Ibrahim, tall and trim in a white *dishdasha*, and Samar, glowing in an indigo abaya, beamed at us. He was an engineer with the Minister of Water Resources, and she was the volunteer co-ordinator for *Humat Dijlah*. They had four children. A young boy and girl tried to melt into the curtains but fifteen-year-old Ahmed, an experimental moustache sprouting on his upper lip, came confidently to shake my hand. Their eldest daughter was away, but we learned she wanted to be a pilot.

Ibrahim, Salman and I settled down in a plush living room, and Emily and Hana went to help Samar finish preparing the *iftar* meal. Ibrahim combined his work at the ministry, where he oversaw all the water projects in Tikrit, with volunteer work as a civil society activist. He'd often help campaigners put together arguments against the actions of his employers. The way he saw it, he could help most by trying to change things from both the inside and out.

He had lived by the Tigris his whole life. His family were farmers, from a village south of Tikrit, and as a child he swam across the river to an island to cut firewood. They lashed the logs together into a type of *kelek*, he said, to float back. One of Ibrahim's passions was charting the usage of boats on the Tigris. Here in Tikrit the most popular had been a circular boat, rather like a coracle, called a *guffa*. Traditionally these were made with coiled reeds, wound into a basket, with a sturdy branch to maintain the shape of the structure. The smaller *guffas*, for fisherman or private river crossings, were 1.5 metres in diameter, but the cargo *guffa*, which could transport anything from watermelons to wheat to a handful of passengers – some photographs from the early twentieth century show twenty-two passengers – were double the size.

The basket was coated in tar for waterproofing and these vessels, like the *kelek*, had the benefit of being light, high in the water and easily paddled in all directions. The Iraqi-German artist Rashad Salim, who had become a good friend during the planning of this journey, had recreated ten fishing *guffas* with the guidance of a woman in Babylon, who he believed to be the last master boatbuilder. Babylon is also where Herodotus noted seeing them, remarking that they were 'the greatest marvel of all the things in the land after the city [Babylon] itself'.

The *guffa*, or something very similar, even appears on the friezes of the Assyrians as two men row a circular craft on a military expedition. They can be seen regularly in twentieth-century photographs and literature from Iraq. One picture that I remembered showed a cargo *guffa* in Baghdad in 1942 carrying a large shipment of 'Black & White Scotch Whisky'. Rashad, who had been part of the flotilla that had travelled on parts of the Tigris a decade earlier, had found in his research that no traditional *guffas* had been made since 2003.

Ibrahim believed the last *guffa* went out of usage in Tikrit in 1942. I mentioned the whisky picture from Baghdad and he laughed. They lasted another ten years in the capital, he said. In Tikrit, they were used for river crossings and for traders to carry goods up and down the city. After that, a *duba* – an oversized *kelek* – was employed, with oil tanks for buoyancy. It was probably a bit more robust, and the level of skill involved for production was nowhere near as great. But that also ceased in 1980 in Tikrit, though Ibrahim reeled off a list of villages where the *duba* was still important. In some it had actually been revived after ISIS bombed bridges, and people reverted to the heritage of their forefathers.

For Ibrahim, the recent history of the Tigris was a passion. He recalled 1986 as the year of the first pumping station in his village. Before that, donkeys were used to carry the water. But it wasn't until 2004, when a new pumping station was built with the help of the Americans, that supply was consistent.

'It should have been great,' he said of the time after that. 'But people have become so desperate, and the majority of the river is used for artificial things, negative things.' Aggregate mining was the worst. From his work in the ministry, he knew there were 170 mines between Shirqat and Samarra city to the south. Seventy-two of those were in the Shirqat area alone. Only 10 per cent were legal.

'This is what I mean when I say the majority is artificial. The mindset is still from 2004. It's survival. No one thinks about the damage it causes to dig into the river. No one thinks about fish, ecosystem, pollution. We have to slowly change that, and change the government's mind, because they also don't care about protecting the Tigris.'

When the call from the mosque sanctioned it, we ate together, rapidly, messily. Afterwards, Samer told us how she and Ibrahim met at university in Baghdad. She was from the capital, but even when they moved to Tikrit she never felt distant from her city. 'It was because of the river,' she said. As long as they lived on the Tigris, she felt happy. 'It connected me back to my home, so I could never feel alone.'

As a family, they regularly went to the river for fun. They'd cook rice and fish on the banks, and swim together. Sometimes they'd race, said Samar, and once in a while the men would let her win. Ibrahim and Ahmed feigned shock, and swore it was always fair.

Like her husband, Samer tried to educate those around her about the need to protect it. When she heard about *Humat*

Dijlah, and Salman, she applied immediately. 'I thought, if he doesn't pick me as co-ordinator for Tikrit, I'll try to bribe him,' she joked. Later, as we tried to sleep in an airless motel room in Tikrit, Emily whispered to me that she felt a little better about people again.

Chapter Twenty-Three

Malwiya

Days 39, 40
Samarra | Duthuloiya
River miles: 602

We travelled in convoy with the soldiers by road past Saddam's birthplace at the village of Al-Awja on the west bank, and also the site of his capture in the spider hole at Ad-Dawr on the east. Both were controlled by *Hashd*, and it was under their supervision that, fifteen miles south of Tikrit, we re-entered the river north of Samarra city. The central strand of the Tigris was now beset by fissures on all sides, leading into high reeds and soft, marshy ground.

Salman reminded us that this was one of the most important points of the journey. It was here, at this latitude on the Tigris, that the geography of the land changed, and thus too its history. Below Samarra, and its equivalent point at the city of Heet on the Euphrates, the landscape becomes a huge, open floodplain as the grade of the rivers soften. Ancient Mesopotamia could be split in two across this line. To the north, where we had come from, were the areas which could sustain agriculture through naturally occurring precipitation. To the south was an area of roughly ten thousand square miles, which was hot, dry and severely lacking in natural resources. The only source of hope was the river, and it was at Samarra that its power was realised. Around 5500 BC, the inhabitants first dug irrigation trenches and ditches, which channelled river water into the arid land beyond. This development in agriculture allowed settlers to expand into previously barren areas now transformed. It was that movement, seventy-five centuries ago, that led to the emergence of the first cities of the world, in Sumeria.

Samarra was purpose-built as a new capital by the Abbasid Caliph Al-Mu'tasim in the middle of the ninth century. He scoured the land along the Tigris and settled at a spot on the east bank that was good for hunting, which was important to him, and bought it from a Christian monastery. Al-Mu'tasim's city was laid out with the Great Mosque at its centre. Eventually, scores of palaces would be built around it and along the riverbank. On the west side, in richer soil, he ordered the construction of canals to irrigate orchards and gardens. This provided food and boosted the sovereign wealth. Al Mu'tasim knew, like those before him, that mastery of the Tigris was the basis of success, and only with that foundation could he embark on his building sprees.

We travelled in a fifteen-foot boat with two armed *Hashd* soldiers through the marshy margins of the city, past the remains of an Abbasid construction called 'the Lover's Palace'. It displayed many of the hallmarks of the Abbasid style: imposing and rectangular, built over two floors with recessed niches in the exterior walls for sculptures, and tall, cylindrical towers punctuating the walls. Around the outside was the depression where a moat once ran. It was open for tourism, technically, but few people ever visited. The soldier said it seemed boring, and he hadn't visited.

The *Hashd* controlled all this area. They had cameras

rigged in the reeds to detect movement, and if any boats were spotted that weren't operated by their men, they'd shoot to kill. Their territory extended as far as a barrage, or regulator dam, built across the Tigris in Samarra. It had been there since the 1950s and sent floodwater along a canal to an artificial lake called Tharthar. The area was sensitive, said the soldier, because of the strategic value of the dam.

Sensitive was a good word for Samarra. Although the city has a predominately Sunni population, it is home to a shrine called Al-Askari, in which the bodies of the tenth and eleventh Shi'a imams are buried. It dates to the middle of the tenth century and nearby is said to be a tunnel into which disappeared Muhammad Al-Mahdi; the twelfth and final of the imams that one prominent branch of Shi'a believe to be the spiritual successors of the Prophet Muhammad. They expect the Mahdi will return, alongside Jesus, to bring salvation to the faithful and establish heaven on earth.

In 2006, Al-Qaeda bombers entered Al-Askari and destroyed the teardrop-shaped golden dome. Although the site is also sacred to Sunni Muslims, the attack precipitated a slide into sectarianism, with Sunni civilians attacked for the crimes of the terrorists. It's estimated that over a thousand Iraqis were killed in the five days after the bombing. For some, the event should shoulder some of the blame for the ideological chasms that drove increased violence in the following years. In 2007, another attack destroyed the two remaining golden minarets at Al-Askari.

The now rebuilt shrine was protected against ISIS by the military, but it remained a potential flash point and required the utmost safeguarding. Inside the city, all security was handled by a militia called *Saraya Salam*, or the Peace Brigade. Prior to 2008, they were the *Jaysh Al-Mahdi* – the Mahdi army – and were created by the cleric Muqtada Al-Sadr.

They gained recognition through armed opposition to US occupation and disbanded in 2008, but were revived in 2014 to fight ISIS. Now they protect the shrines in Samarra, as well as undertaking limited operations elsewhere in the country. They are still under the leadership of Sadr. During our travels, Sadr was in the news even more than normal. Iraq's parliamentary elections had been due to take place in June but were pushed back to October. This helped us as we were no longer travelling in the immediate slipstream of election tensions. Very few people we met had any intention of voting, but Sadr was the head of one of the biggest political blocs. His image was everywhere, and we'd find more supporters as we headed south.

One way in which the distinction between the *Hashd* here and the *Saraya Salam* militia ahead was clear could be seen when we docked close to the regulator dam. Here, the *Hashd* had a small jetty and, beside it, a thatched hut made of reeds. This was in the style of a traditional *mudhif*, which came from the marshes in the south, and historically had no place here. It had been constructed symbolically to affirm southern, Shia, stewardship of this area.

Inside, much like a shrine itself, were posters of the spiritual, political and military leaders that this group revered. On the right was Imam Hussein, and on the left was Grand Ayatollah Ali Sistani. In the centre, largest of all, were recent martyrs: Abu Mahdi Al-Muhandis, kneeling stoically behind sandbags and, clasping his hand paternally, Iranian general Qassim Soleimani. At the top was Sayyid Ali Khamenei, the Supreme Leader of Iran. It was a not-so-subtle sign that this brigade was aligned with Iran. Inside the city, *Saraya Salam* and Sadr had vociferously moved away from Iranian interference in Iraqi affairs in recent years. As we stood there, with the Iran-influenced *Hashd*, preparing to meet Federal soldiers to negotiate permissions, then to pass into an area controlled by *Saraya Salam*, in a city populated by Sunni civilians, the fragmentation of Iraq seemed greater than ever.

What got lost in all this was the importance of the wet-lands. They existed because sedimentation had built up behind the barrage, and the reeds and submerged aquatic vegetation that grew there in turn attracted significant num-bers of migratory wildfowl and raptors. Iraq is an important habitat for bird migration, despite recent instability. Since we began we had seen not just permanent breeding species, but also those overwintering. Soon we would see others that came for the summer.

Here in the wetlands, benefiting from the moratorium of human activity, the vulnerable marbled duck found a habitat, as did the Iraq babbler. I was certain we saw one, with a long tail and streaked back, but it was too far away to be sure. It didn't matter. The joy was in wondering. There were over seventy bird species here, said Salman, and probably seven or eight fish species, but it had been hard for specialists to conduct surveys. And, despite the restrictions, there were still determined hunters who found a way. The water was also vulnerable to oil spillages from Baiji, which was true for this entire section of river.

Captain Saif and his men met us at the regulator dam. No one could decide where was safe to sleep, so we all stayed there. In a pokey office beside the barrage, twelve soldiers and our team of five arranged ourselves like sardines on the floors across two rooms. In the morning I was woken by Tarik, the swimmer, who told me that the chief engineer would like to start work and was waiting outside. I hadn't realised that I was sleeping under his desk, so I apologised, put on my trousers and left the room.

I hoped the chief engineer might be interesting to speak

to, but the man was six months away from retirement, and he answered my first question about the dam by telling me the water was clean, pollution didn't matter and that he was sure dams had no downsides. We left him to his papers and retirement planning. Outside, by the sluice gates, Salman, Tarik and I made scrambled eggs for the team, which we then ate crouched on our haunches, scooping out lumps from the pan with *samoun* bread. Captain Saif mentioned rather casually that he had gone to primary school with Abu Bakr Al Baghdadi, the leader of ISIS. This seemed an amazing piece of information to have held on to for so long, especially given how many times I'd heard the story of his gastric bypass.

'They were a very poor family,' he said. 'And very religious. He was a calm, peaceful boy. One time he got hit by a bully, and he just said, do it again, and he stood there. I don't know what happened.'

The soldiers insisted on coming with us for the day, under the auspices of security, but Hamoud, the singer, also said quietly that they were excited about visiting the Malwiya minaret. It was the site I had been most looking forward to anywhere in Iraq. We approached it from the south and soon saw the distinctive spiral, starting on a square foundation and then curling up over six levels, each narrower than the last, until it topped out on a small cylindrical platform 170 feet in the air. Beside it was the rectangular wall of the Great Mosque of Samarra, which had been mostly destroyed by the Mongol warrior Hulagu, a grandson of Ghenghis Khan, in 1258. We would see more devastation from that invasion downstream.

The mosque and minaret were built eleven and a half centuries ago by Caliph Al-Mutawakkil, son of Al Mu'tasim. Al-Mutawakkil became known widely for his magnificent building works, but also for his excess. He embarked on ambitious canal projects to bring water to new palaces along the Tigris, and even built another new city, eighteen miles to the north of Samarra. It is said he was a drunkard, and his reign marked by religious intolerance, during which he forced Christians and Jews to wear cloth markers to distinguish them from the Muslims. He neglected the irrigation ditches that watered agricultural areas, instead focusing on vanity projects. His squandering and division eventually led to his assassination, partly planned by his son, and in the end caused the decline of the Abbasid Empire, which never recovered from weaknesses he created.

A solitary guard, bewildered by our motley group, tentatively took a small fee from the five of us, and waved the soldiers through. Immediately, they tore off to climb the minaret. The sky was filled with migrating swallows, falling on us as we chased the soldiers. The walkway was sandstone, like the rest of the structure, and there was no guardrail. Instead, one walked in circles, following the snail shell, squeezing into the wall if someone came down. As I climbed, a family of four staggered towards me, the mother's feet so close to the edge that her abaya trailed off into the void.

On the way up I overtook all the soldiers but one. Three sat down halfway, exhausted, to light cigarettes. The others worked their way up manfully, but slowly, with a lot of heavy breathing and some swearing that didn't feel appropriate for a minaret. When I summited, it was just me and Karar, a young, slim soldier with a perpetual grin on his face. The ruins of the Great Mosque were laid out below like a schematic. Swallows dived around our heads, and beyond the city slouched out into cracked earth.

Once Samarra had been arranged by district, with each area reserved for a distinct craft. Now it had grown, as most Iraqi settlements had, on the basis of a million small-scale decisions. The area was a UNESCO World Heritage Site, which at least had prevented the construction of high-rise apartment blocks. But the rest had developed according to the needs of each resident or street rather than any top-down urban planning. In the centre was the golden dome of Al-Askari, its colour dulled by the gloom of the sky. Salman arrived, a little breathless too, and smiled.

'People say the design was based on a ziggurat, with the different layers,' he said. This was what the 1982 *Tourist Guide* suggested, too.

'What do you think?' I asked.

'I think it was designed for exercise.' He sat down, with me and Karar beside him, and together we waited, enjoying a moment of respite, at the top of a tower, where no worries could keep up.

The river bore us south-east as we left Samarra and date palms crowded the reed beds. Already we could feel a greater ease with which the river meandered. At one particularly tortuous curve, where the Tigris made a full U-turn, we docked at a riverine town nestled into this digit of land. It was called Duthuloiya, and the fisherman who brought us from Samarra seemed perplexed when we asked if anyone used the river. Couldn't we see, he asked?

Around us the Tigris was alive. A dozen children swam by a jetty, diving and thrashing. In both directions, there were private docks with expensive fibreglass-hulled boats. Then, behind us, a bright red jet ski bounced by. The man aboard hit the throttle and took off upstream. Distress waves were sent to shore as the river recoiled. The jet ski pulled a tight turn by some reeds under an olive grove on the bank above. Then it sped back, carving a donut in the middle of the Tigris. A little farther on, in a quieter meander, an eccentric artist had built a two-storey paddleboat capable of carrying a 150-tonne load. It was a remarkable piece of innovation, built entirely to his own bespoke specifications, inspired by a Mississippi paddle steamer.

On the corniche there was the usual guy in a polo shirt and hat, with the usual bulge in his belt. To my surprise, this time the *mukhabarat* really did just want to check in. Duthuloiya was safe, he said. But would we like to join him for *iftar*? We ate together at the home of a retired major general who thought at length before he answered questions. There were around a dozen guests at his home that night, and every night during Ramadan he hosted different friends from the town. We ate outdoors, on the grass, in a courtyard lined with pomegranate and orange trees.

'When you have a tree full of fruits, you throw a rock at it,' he said. 'This is a saying here. I think Iraq is like that tree. Everyone wants its fruits, and they're willing to attack it.'

He never liked wars, he said, though he'd been involved in so many. He predicted the next war would be over water. 'It's certain. Since 2010, Turkey hasn't honoured their water sharing agreement with Syria. Look what's happening there,' he told us. He referenced a verse in the Quran which said everyone should have access to water. 'Turkey is a Muslim country, so they should feel bound to share. But they don't, so this is what's in our future.'

One of the other dinner guests was an old man wearing sunglasses. When he took these off, I saw his face was contorted around permanently closed eyes as if the muscles had all drawn in around the lost sight. He had studied as a mullah for twenty-five years, and he wanted to talk to us about swimming. 'In Islam, we say teach swimming before writing,' he told me. 'Because you can always find someone to read and write for you, but not to swim for you.'

He was twelve when he went blind. He called it the black water, flooding his eyes. His abiding visual memories of the world were of the Tigris, when he swam back and forth as a boy. 'I only lost one sense. I still have the others,' he said. 'We blind people can feel everything. We see from our hearts. We see more than most.' He knew the river was sick, he said. He could hear it. In the last ten years, the sounds of the foxes had disappeared. The river was unsettled, and its voice weaker.

The moon appeared, and in a cloudless sky I lay back to look for constellations. Emily rested alongside me, her cameras nestled under an arm. By this point we could sleep anywhere, anytime. At some point blankets were brought, guests drifted home, the fire settled into its embers, and the sound of the river carried across quiet streets to wash over us.

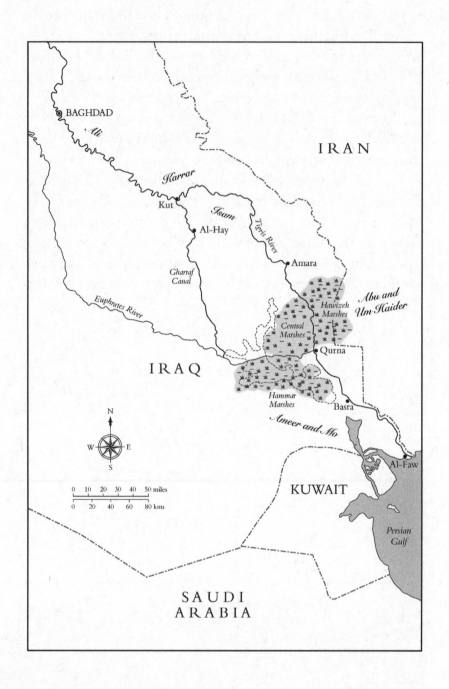

PART THREE

LOWER TIGRIS

BAGHDAD TO THE PERSIAN GULF

I greet you from afar, O greet me back,
O blessed Tigris, river of gardens green.
I greet your banks, seeking to quench my thirst,
Like doves, between water and clay aflutter seen.
O blessed Tigris, oft have I been forced to leave.
To drink from springs which did not my thirst relieve.

Muhammad Mahdi Al-Jawahiri,
translated by Hussein Hadawi

Be he philosopher or teacher, historian or poet, lawyer or
reformer, modern man will find his prototype and counterpart
in ancient Sumer.

Samuel Noah Kramer, *The Sumerians*

By far the most serious long-term threat the country faces
is from the potential economic impact and environmental
devastation of climate change. According to the UN
Environment Program, Iraq is the 5th most vulnerable country
in the world to the consequences of changes in the climate.

Barham Salih, President of Iraq 2018–2022

Chapter Twenty-Four

A Model of City Planning

Days 41–45
Baghdad
River miles: 681

We drove to Baghdad through a tribal area called Tarmiya. ISIS were active there, and four members of a Sunni tribe had also been killed that day in a gunfight with some *Hashd*. Salman said it was possibly the most dangerous area anywhere on the river. But in the same breath he quietly confided that he hoped this would be the last time we detoured around a risk area. South of Baghdad we didn't need to worry about jihadists, and he was optimistic we could manage the militias. It was a new start for us, he said. 'Now we can just focus on nice things,' he grinned, the way he always did when he was being a little facetious. 'We'll go and look at Baghdad's art and music, and we don't have to worry about guns and permissions. It'll be like a holiday.' Soon he was giggling, and so were we. I adored him for his boundless optimism as much as his silly humour.

Baghdad has long been associated with culture. The city sprang up on the banks of the Tigris in the eighth century to be the capital of the Abbasid Empire, led by Caliph Abu Ja'far

Al–Mansur. It became, for a few hundred years, the centre of the world, and exemplified the Abbasid ambition. Goods from East Africa, India and China were transported by river to furnish the burgeoning city. The innovation and manufacturing of paper helped Baghdad write its own place in history as a hub for learning, and soon the capital was known for its pre-eminent poets and philosophers; scholars and scientists; architects and astronomers.

Al–Mansur designed a great circular city with outer walls eighty feet high and four grand gates that shepherded arrow-straight roads to meet in the centre. ('A model of city planning,' reads the 1982 *Tourist Guide*, 'and one of the great centres of civilisation.') The location benefited not just from one river, but two: the Euphrates comes within thirty miles of the city, and is linked to the Tigris via a series of canals.

Time hasn't been kind to the aesthetics of Al–Mansur's city. In 1258 Baghdad had been sacked by Hulagu. Thousands were massacred, and the Caliph trampled to death. The grand library was destroyed, and it's said that so many books were thrown into the Tigris that the water ran black with ink. This brought an end to the Abbasid caliphate, and perhaps too Baghdad's glory days as the centre of the civilised world. The foundation of its power – the canals and irrigation systems that harnessed the river – were also destroyed.

In 1327, Ibn Battuta paused in Baghdad, waiting for a caravan to leave to Mecca for Hajj. He lamented the decline of the city and seemed only to find joy in the many bridges across the Tigris, and the constant flow of people from one bank to the other. He was bemused by the fact that Baghdadis took such pleasure from simply walking across such a significant body of water.

A century and a half earlier, however, even before the Mongol rampage, the Andalusian traveller Ibn Jubayr also wrote that the true splendour of the city was found in the river:

> There is no beauty in Baghdad that arrests the eye, or summons the busy passer-by to forget his business and to gaze – except the Tigris, which lies between her eastern and western quarters like a mirror set off between two panels, or a necklace ranged between two breasts; she goes down to drink of it and never suffers thirst, and views herself by it in a polished mirror that never suffers rust; and between her air and her water feminine beauty is brought to its flowering.

It would be hard to disagree with either traveller today, at least on first appearances. Now, under a smoggy sky, the city appears as a haphazard jumble of highways and intersecting neighbourhoods. Decades of conflict have damaged much of the infrastructure. There are checkpoints and concrete blast walls along major thoroughfares, though fewer than before. It is a city more used to looking behind than ahead. Still, despite everything, the Tigris continues to cleave the city in two with grace and poise, and today, just as in 1327, it is a wonderful thing to stand on a bridge looking down over the parted city.

We were still travelling during a pandemic, but Iraqis had by and large long given up on taking precautions. We rarely saw any masks and life carried on as normal. Most people could not afford lockdowns or other restrictions. We continued to wear masks and test regularly, even as anomalies, and one reason we slept in tents so often was to avoid being in close contact with others.

In Baghdad, it came as a surprise to find empty streets. There had been a spike in cases and the government implemented a sudden weekend lockdown. There'd been rumours in previous weeks, but no action transpired. Now, to everyone's surprise, it was enforced. There were some pedestrians here and there, but cars were banned entirely. The prevalence of military became even clearer to see, no longer diluted by the traffic of a city of eight million. Every major intersection had some sort of rudimentary roadblock, and only blacked-out SUVs driven by the powerful and elite could pass through.

With a yellow heat lying on the city like a weighted blanket, we walked along the eastern bank of the Tigris and into the park named for the poet Abu Nawas. The grass was balding, struggling in the sun and worn thin by picnics, and concrete paths criss-crossed between topiaried bushes. Under gently swaying palm trees, a few young couples in counterfeit Adidas trainers and Gucci jackets squeezed together for selfies, channelling, probably unknowingly, some faint memory of the poet.*

The river path led eventually to Al-Rashid Street, where fading Ottoman-era buildings slowly collapsed into asymmetry, then past ornate façades of British-era cinemas, now hardware stores. Mutanabbi Street, which connects Al-Rashid to the river, has been a hub for Baghdad's artists and academics for almost ninety years. The narrow, pedestrianised street is verged by bookstores along its length, which on Fridays display their wares on the cobbles. It is still a manifestation of the saying: Cairo writes, Beirut prints, Baghdad reads. Coffee shops fill with poets and bibliophiles, and last time I was there a young Iraqi busker in a leather jacket was playing 'Folsom Prison Blues' by Johnny Cash.

But this Friday shops were shuttered, and a pack of dogs chased each other around a pile of cardboard. The only

---

* In the time of the Abbasids, Abu Nawas was a pioneer of *ghazal* and *Khamriyyat*, roughly translated as the poetry of wine and of love.

person there was Ali Al-Kharki, who had come to join us. Ali was the other co-founder of *Humat Dijlah*, and we knew him well. He was young and suave and wore a white linen shirt. When he spoke, he had that natural authority that some command. He and Salman were the perfect team. Salman was everyone's best friend. Ali made you listen. Together, we walked back through empty streets, occasionally remarking that we'd probably never see its like again.

In the middle of Tahrir Square, close to Al-Rashid Street, is the Freedom Monument, one of the best-loved works of art in Baghdad. It is 150 feet long, 30 high, and shows figures in bas relief playing out the events that led to Iraq's moment of independence under Abd al-Karim Qasim in 1958. In October 2019, Tahrir square – which is actually a large roundabout – was the centre of a mass protest against the Iraqi government. Mostly those calling for change were the young and disenfranchised, and they gathered to demand the resignation of the prime minister, Adil Abdul-Mahdi. The protesters were peaceful and set up tents around Tahrir. Many of Emily and my friends from Baghdad lived there for weeks at a time. Ali was one of those.

The protests were met with brutality by the Iraqi government. Tear gas and live ammunition were deployed alongside water cannons, and social media at the time was filled with disturbing videos of injured and dead protestors. For a while, the internet in the country was shut down. Snipers were positioned around the square, and the *Hashd al-Shaabi* blamed for some of the worst of the atrocities. In one two-day period alone, 107 civilians were killed and 2,458 injured.

The Freedom Monument was initially deserted when we arrived until, from the far side, came the clashing of batons on shields and a mechanical siren. A battalion of riot police appeared in formation. Behind them a navy-blue Humvee fired jets of hot water out onto the street. 'They're training,' said Ali, noticing my confusion. The lockdown offered a rare opportunity to practise their drills in situ.

As the men trotted past, some smiled and winked at Claudio's camera. It was all faintly absurd. One policeman stopped to adjust his boots, and the man behind him tripped over. The trainee water cannon operator soaked his own men. The marching was comically mistimed. But for Ali and Salman, there was no joy in the farce. They had lost friends at this square. Both had to leave their homes because of what happened here. Iraqi youth came out to the streets to challenge corruption, and lack of opportunities, and foreign interference in government, and to ask for a new political system, Ali said. And what did they get as an answer? Shot at, beaten and vilified.

Salman's family asked him not to stay in Tahrir during the uprising and he respected their wishes, despite desperately wanting to be there. Instead, he visited and offered support. He began by sharing information on Telegram groups for protestors – updates on threats, advice on how to stay safe. Every couple of days, he went to the square and gave presentations at a tent for environmental activists. He helped with cleaning up the rubbish each evening. Ali stayed at Tahrir for weeks, but Salman always returned home at night.

On 11 December 2019 it was raining, and Salman went to buy a tent for those sleeping on the square. He was travelling back from the shop with his friend Omar, heading for the Spanish Embassy for a meeting, when their car was stopped at a makeshift checkpoint. The *mukhabarat* emptied their bags and found Salman's black and white scarf. They were looking for anyone involved in the protests and this, he said, was a giveaway.

Salman and Omar's phones and laptops were immediately

taken, before they could contact anyone, and they were taken away. That was the last anyone heard from them for seven days. Even now, Salman can't talk of this time, except perhaps in his sleep. Throughout our journey, I had learned bits and pieces about his experience. He'd had to have an operation on his left eye because of an injury sustained by the beatings. When, during his detention, international organisations and journalists made statements demanding his release, these were read out to him and he was tortured more as a punishment. Occasionally, when he was telling me about something in his life, he'd pause and rub his temples, and realise that the memory was gone. He'd lost a lot since 2019, he said.

He was eventually released, in large part because of his public profile and the pressure the international community had put on his kidnappers. On 22 December, he left for Kurdistan with his wife and daughter. A week later, Ali went, too. Other coordinators and civil society activists were being targeted, and many disappeared. It was our great fortune to know these two men, who had spent their adult lives demanding peaceful change in their country, environmentally and politically, and as a result were living in exile in the north, always watching their backs. The irony now was that Salman's kidnappers were connected to the same *mukhabarat* who helped us along the Tigris. All of them knew his story, and although he was no longer at risk from them, it kept the trauma close to the surface.

Close to the square was an underpass filled with graffiti and artwork from the protestors. One, on a sky-blue background, showed a bullet passing through the UN logo. '*Where are you?*' it demanded of them. Beside it, a painting of a young boy looking out, a tear-gas canister about to blast through the side of his skull. Other images used the Iraqi flag, showing crowds draped in it, or fingers writing in blood that seeped out and bound the black to the red. Those demanding change were patriotic, and proud to be Iraqi, said one. It was the government and system that they had a problem with.

'These are the martyrs of the revolution,' said Ali, pointing to portraits of two men who had been killed early on. The underpass became a space for artists to express themselves and showcase their skills. This revolution, Ali said, was sixteen years in the making. To him, 2019 represented a greater shift than 2003. This was the point at which youth took initiative and claimed a platform.

I asked what made the authorities react so violently. It was fear, he said. All over Iraq, protestors went out into city squares, just like in Baghdad. At least seven hundred were killed in less than six months, but still they remained. It was only the pandemic that finally cleared Tahrir. Now the municipality wanted to remove the murals, but Ali felt confident. 'The existence of these drawings keeps the idea of the revolution alive,' he told me. 'No one will be able to remove them. It's part of our history. There's not enough power to stand against the Iraqi people who want this.'

Across the street, a brick wall listed the names of all those killed, written in white over a black background. Rows and rows of tight, neat Arabic script, each representing a life lost in pursuing better governance. Emily took a picture and immediately two police officers arrived. Ali spoke politely and calmly, and I couldn't help but admire his composure. Once again, I felt privileged to be in the company of someone who, to us, represented the positive future of this country.

Chapter Twenty-Five

Bridge to Bridge

Days 46, 47
Diyala River confluence | Taq Kasra
River miles: 699

We had arranged access to travel through Baghdad by boat. This was one of our more notable achievements, though I could claim little direct credit. Emily had convinced a contact of hers to put us in touch with someone influential, and, over the course of a few months, a plan took shape. In the city, river permissions are allocated according to bridges. A boat might be allowed to move between Jumariyah Bridge, opposite Tahrir Square, and Sinak Bridge, but cannot travel beyond that quarter-mile stretch. To go the whole way was a rare thing, if not unheard of. We were granted passage in a police boat, with two soldiers and a member of the *mukhabarat*, all the way through the city, and they would tell us what we could and couldn't photograph.

We began at a *duba* bridge, beside an artificial island filled with holiday chalets. Large villas on the banks, cloaked in palms, fronted onto the river. We had the broad, languid highway of the Tigris to ourselves and I thought of Al-Mansur being rowed up and down, looking out across the scrub either side, trying to imagine his new city. As we approached the centre, Ali pointed out two blocky buildings. One was the Ministry of Environment and Health, and the other Baghdad's largest hospital. The water by the bank was a murky grey, and a film of scum floated on the surface. 'This is the worst pollution in the city,' said Ali, raising his voice. The area had long been an irritation for *Humat Dijlah*. The hospital, called Medical City, was sending its wastewater directly into the river. 'It's blood and body parts, and chemicals from the operations. You can smell it, right?'

There used to be a treatment station, said Ali. 'But it hasn't worked since 2003. What we're smelling is the waste of these processes in the hospital. It goes directly to the river. And people drink this.'

We pulled the boats to the bank to look for the pipes. An isthmus of land had been built in front to hide them, and we had to climb up and over. There, grey, foaming water poured out of a grated opening six feet high. A few scrubby plants hung on at the side, black and drooping. The toxic pool whirled, and smaller pipes below drained it on into the river. Some were hidden under the surface of the Tigris and perhaps all were meant to be, but at these lower summer water levels there was no disguising them.

'This area is highly secure,' Ali told us as we stood by the poisoned pond. 'It's because they don't want anyone to see what's happening.' It was guarded from the road, and normally no one could access by boat. Water-quality testing was banned, Ali said, though *Humat Dijlah* had carried out an investigation in 2018. Their results were unsurprising. People were getting sick. They were being poisoned by the water, and the report suggested a huge spike in cancer cases in the neighbourhoods around this area. As usual, it was almost impossible to tie these directly to any single source of pollution. 'The joke is that it's happening directly in front of the Ministry of Environment and Health,' Ali said, shaking his head. If it could happen here, so unashamedly, what hope did the rest of the country have?

Emily finished photographing the pipes and asked why nothing was being done. 'Our analyses suggest the problem is too big,' Ali said. 'Whenever ministers come in, for a short time, in an unstable political environment, they're not interested in getting involved with a deep project like this.' *Humat Dijlah* had solutions, he said, and a short- and long-term plan. But no one took them seriously.

Soon the outlines of buildings by the water were burned into hazy silhouettes by heat and smog. Some of the nearby shapes were famous hotels, where top floors boasted river views. There was one lone swimmer, who I both pitied and admired, and sporadically a twisted metal protrusion of some dumped vehicle or piece of infrastructure. The *mukhabarat* told us here and there to avert cameras from government buildings, but even our gaze was returned with scrutiny by stony soldiers in sentry points. Theirs could not have been an interesting job.

A dozen wooden fishing boats, the first we'd seen in Baghdad, were tied to a bank adjacent to Al-Rashid Street. Each was limited to a quarter-mile stretch of water. Most had a canvas covering to protect from the sun and also took tourists across the river for a small fee. In 2018, said Ali, the water level was so low that these fishermen could walk across the river without their feet leaving the bottom.*

Opposite stood a man waist-deep in the water. His name was Adil, and he was an Egyptian gold panner. For forty years, he said, he'd been coming to the Tigris, from 8 a.m.

---

* In summer 2022, young men played volleyball in the centre of the Tigris in Baghdad and the river level barely reached their waists.

to 4 p.m., six days a week, with buckets and sieves. He wore no special clothing except two soggy work gloves, and was alone, save for a shy teenage assistant who retired further up the bank when we arrived. There used to be dozens of goldsmiths working in stores above the river here, he said. Anyone in Baghdad who wanted gold, for weddings or celebrations or any of the other reasons people like gold, would come to this part of the city. And, every morning before opening, goldsmiths trudged down to wash their equipment in the Tigris. 'Gold comes from nature,' said Adil. 'So it'll always try to go back there. I've made a living getting the bits that escaped.' The dust and shaving that got washed away kept him in business.

In the 1980s, Adil had a team of thirty-five working along the bank. But in 1998 the city remodelled the corniche and, slowly the gold workshops and stores disappeared. Now everyone buys gold from the Gulf countries, Adil said, and local craft workers are gone. No one comes to wash gold in the Tigris any more. No one comes at all, except Adil.

We watched him plunge a large red bucket into the silt and then swirl it, slowly tipping out water. Once most excess was gone, the sediment was split and swilled again until individual grains could be separated out. If anything glinted in the light, Adil might be in luck. These days he usually found around a gram a day. It was 1 per cent of what he found before, but enough to live on. 'I'm not going to get rich,' he said. 'I'm here because I like it, and I'm good at it, and what else would I do?'

South of Jumariyah Bridge we came to the Green Zone, where politicians and diplomats were holed up behind blast walls, and we drifted alongside villas with fences of razor wire that Salman said were used for fancy parties. Other buildings, nondescript, were under the control of the militias, and these were the ones we had to be most careful of. Even with the camera off, we were told to look the other way, too, as if in the presence of some strange, blinding greatness.

The river that once bound two sides of a city together, and connected north and south, had been cleared of all transport, trade, culture and life. Now it was a boundary. The demise of the river affected people in many ways, Ali said. During the sectarian fighting in Iraq after 2003, the Tigris had become a dumping ground for bodies killed in the conflict. The smell of decomposing corpses in the water hung in the air, so that everyone in the city knew it and feared it. Now the city was peaceful, relatively speaking, but those who lived around Medical City recognised the smell from the river. That triggered trauma and memories of that time. Just imagine, Ali implored me, but I could not. There were too many layers of sadness in that thought for me ever to fully grasp it.

'In our methodology, we say that pollution is like terrorism, but even worse,' said Ali. 'Terrorism targets a group of people. Pollution targets millions. We tell our volunteers and people in our campaigns that pollution is the worst, and that's why we call on the government to take care of its people. They fight against terrorists, but they should fight against this.'

Just as the city slips away, another major tributary joins the Tigris from the east. This is the Diyala River and, like the Lower Zab, it comes from Iran. This confluence had been the site of at least two major battles: one between the Neo-Assyrians and the Elamites from modern-day Persia, and the other in 1917 when the British defeated the Ottomans and took Baghdad.

We detoured for five or six miles upstream and saw only a dying river, far from the epic staging ground of imperial warfare. There was barely enough water to reach the Tigris, and what remained hardly moved. It looked stagnant and could certainly not carry a boat. Green sewage floated on top. We walked down to the bank beside a farmhouse, covering our noses with scarves. A farmer in a long white *dishdasha* said it had been getting worse every year. A decade ago they drank from it. Now, it even poisoned their animals, so he had to buy drinking water for his animals.

The bank extended into the middle of the river in a causeway of cracked earth. Those islands of fractured, dry dirt near the farm were stable, but closer to the water they became like a bog. Twice the farmer's kids had been sucked in, and were lucky to be pulled out. Subsequently, they went to hospital with skin infections. His family lived by the river, he said, and he had to ban his children from playing near the water. What madness was that? He'd heard Tehran was cutting the water flow to the Diyala, and it was true that a dam was under construction to divert water away from its natural course and direct it into south-western Iran. Because the volume coming to Iraq was lower, local authorities in Diyala Governorate often failed to release any fresh water at all to these final stages of the river.* The family here was planning a way out of this hell. They'd go anywhere but here. As we left, Emily took a photograph of the man standing alone by the stagnant water, head and shoulders slouched in defeat.

We slept in the garden of a sheikh, surrounded by groves of date palm, and by the time we left the next day, it was already 40°C. Our nights in the tents were becoming sticky and fitful. In the mornings there was usually a tap

---

* Again, in 2022, as elsewhere, things got worse. Due to water shortages the government cut Iraq's cultivated areas in half, and no crops were grown in Diyala Governorate. The river became even more stagnant than when we travelled.

or a bucket to wash from, and an outhouse or a bush that provided adequate cover. We applied thick, zinc-heavy sun cream, and filled a cool box with ice, water and juice from village stores. We were spending ten to fifteen hours a day under the sun, and easily drank five litres each. Heatstroke was a concern.

On the west bank, three miles back from the Tigris, was the site of the ancient city of Seleucia. Its glory days had come when the Greeks ruled Mesopotamia, in the final three centuries BC, and in its prime it was regarded as highly as Alexandria on the Mediterranean. On the river's east side was the grand arch of a Persian city from somewhere between the third and sixth centuries AD. The city had been Ctesiphon and the arch, which is still visible, Taq Kasra. Much may still be under the ground, unexcavated.

I had been looking forward to visiting. In general, south of Baghdad, the most important historical sites lie along either the present or ancient course of the Euphrates. We had now entered a new, deeper time period. Broadly, the area between Baghdad and the Gulf, bounded by Iran to the east and the desert to the west, was once known as Sumer. The name itself is taken from another language, Akkadian, which came from further north in Mesopotamia, and means 'the land of the civilised kings'. Around 4000 BC, the disparate communities that lived along the watercourses began to coalesce. One theory is that a dramatic environmental change forced this; the climate became drier, and the rains that settlers relied upon no longer came flooding down the rivers. Larger groupings protected against such risks, with storehouses and a greater capacity for managing large-scale irrigation. But it required a huge amount of work throughout the year to dig out silt from the canals and use it to pack levees along the banks. The Euphrates was shallower than the Tigris, and easier to irrigate from, and so the larger settlements grew at strategic points on the southern Euphrates.

These became the world's first cities. Gwendolyn Leick writes that the Sumerians themselves noted Eridu as the 'Mesopotamian Eden, the place of creation'. She relays how the Babylonian god Marduk is said to have created the world:

> All the lands were sea. The spring in the sea was a water pipe. Then Eridu was made, Esagila was built, Esagila whose foundations Lugaldukuga laid within the Apsu ... The gods, the Annunnaki he created equal. The holy city, the dwelling of their hearts' delight, they call it solemnly. Marduk constructed a reed frame on the face of the waters. He created dirt and poured it out by the reed frame. In order to settle the gods in the dwelling of [their] hearts' delight, He created mankind.

The first city as described by the Sumerians, Eridu, sounds much like the culture of the marsh Arabs, who we would meet before long. But historians generally consider Uruk as the first proper city, and a Cradle of Civilisation. Uruk was home to the legendary King Gilgamesh, whose story was unearthed in the library of Ashurbanipal in Nineveh. In the epic, Gilgamesh leaves home in search of eternal life, battles great beasts and has wild adventures. It caused a stir upon its translation in the 1870s because of its detailed telling of a flood myth that was very similar to, and which massively predated, the biblical flood narrative.

Gilgamesh visits a man called Utanapishtim, who tells him that the god Enlil was irritated by the noise of the world's populations, so sent a flood to destroy them. Ea, the god of wisdom, gives Utanapishtim instructions on how to build a boat and what to put on it, including every animal in existence. He, and his animals, are the only survivors.* But Gilgamesh's Uruk was not the only city at the time. There were more than a dozen others – eighteen, perhaps – each with their own gods, temples, cultivated lands and trade networks. They were constantly fighting one another, and the cuneiform records show that, often, the conflict was over water and arable land.

Later, towards the end of the second millennium BC, Sargon, who may have once been the king's gardener, took the throne and extended his rule over all of Sumer, and much of northern Mesopotamia. This became the Akkadian Empire. Later still, Hammurabi, from a small town within this empire called Babylon, founded a new empire in the nineteenth century BC. The Babylonians, like those before them, clashed with the Assyrians, and vied for control over Mesopotamia with several other peoples from Iran to the east and the mountains to the west. These, however, were histories we would not see up close. The deep channel of the lower Tigris deterred early settlers on the alluvial plain, and instead we would move alongside a more recent past, like Seleucia and Taq Kasra.

Both sides of Taq Kasra were controlled by the *Hashd* and someone in the nearby village suggested they were Iranian-backed factions. This seemed plausible, because the site was an important part of Persian history. Salman negotiated a visit for us, as long as we promised to be quick. The remains were among the most exciting I saw anywhere in Iraq. Two parts of the palace wall are still in place, with huge, vaulted doorways and rows of symmetrical, pillared ornamentation. In the middle is the mud-brick arch, 121 feet high and 85 feet

---

* The Assyriologist who discovered this, George Smith, was said to have got so excited when he finished the translation that he took off his clothes and danced around his room in the British Museum.

wide, launching itself across space in a great curve as a portico for a central hall, or perhaps a throne room. It was 'the
largest single-span brick-built arch in the world,' said my
1982 *Tourist Guide*; an architectural marvel, still seeming to
defy gravity.

We could see no one maintaining the site. The two armed
guards by the wire fence that surrounded it were tasked only
with stopping visitors. A few months earlier, a chunk of the
arch had collapsed in heavy rain. There were now fears the
rest would crumble, too. I'd read that the ALIPH Foundation
was funding some emergency rehabilitation work. There
were also rumours about Iranian investors offering funding.
But on the ground, nothing had visibly begun, and much
of the rubble from the collapsed mud-brick was still strewn
across the throne room.

We sat for a while together, leaning back on the mud-brick
walls, staring up at the arch. It was cool under cover, and
a wave of happiness washed over me. During the break in
Baghdad, I'd spoken to a friend about the challenges of the
journey so far and he reminded me of the importance of joy
in travel. It was the only thing that would carry us through
to the end, he said, so we must take it where we could. Here,
in the winter capital, with cold orange juice and melting
chocolate, we did what we could to embrace it.

Chapter Twenty-Six

Wireat

Days 48–50
Suwayrah | Aziziyah | Numaniyah
River miles: 825

At a village called Suwayrah, a fisherman cleaning his nets by a riverside park beckoned us over. He fancied himself a local historian and he was collecting stories from the old people. Under the Ottomans, an agreement had been made to plant trees and encourage farming, but the pashas were constantly concerned with flooding. Until the creation of Tharthar lake, each year meant an anxious wait to see what the rains would bring. As we left, Ali asked Emily and me, 'Do you see how historically the bigger problem was floods? Now we're seeing drought, and people are confused. They're relying on the same systems of dams and barrages and irrigation methods and hoping it'll solve the opposite problem.' In general, he said, this was one of the biggest issues for Iraq. The infrastructure was designed for flooding, and had never been updated. It was no wonder they were struggling.

We were squeezed into a simple fishing boat, meandering past tamarisk and seepweed which pointed like signposts in our direction of travel. An abandoned grain silo was followed by a long stretch of riparian farmland. Potatoes and watermelon were watched over by young boys who slept and fasted and swam, and made sure thieves didn't steal their cattle. When the heat became too much, we looked for a farmhouse and were usually invited to sit in the shade of a fruit tree. We didn't accept offers of cold water, in case it was unfiltered, but if we were lucky there'd be juice, and when that happened we'd linger longer. Hana was good at finding baby animals, like puppies and tiny, bleating goats, and we passed a few pleasant hours dozing under trees with furry company.

When Emily and I had been planning the journey, Rashad Salim, the artist who had travelled on the Tigris, warned us of wires and rope strung across the river. They were a nightmare, he said. But we had reached Baghdad without seeing any and I presumed he was exaggerating. Now, though, we finally entered the land of the wire. Every farm owner who wanted to cross the river would set a steel cable across the length of the Tigris, two or three inches in diameter, with a small wooden fishing boat attached. This allowed people to row across without getting pulled downstream and it seemed everyone had their own – if there were two homes, there were two wires.

For us, seemingly the only vessel going along the river as nature intended, these wires were a pain. They were strung at irregular heights and sagged in the middle. It was rare we could make it underneath in the central channel. Mostly we scraped in close to the banks, where spiders and insects delighted in the disturbance, and lifted the cable above our heads. The only thing I liked about the wires was that the English word was also used in Arabic. When it became plural, it was conjugated in the Arabic fashion: one wire, two *wireat*. A lot of words that were borrowed from English followed the same rule to pluralise. One mobile (phone), two *mobileat*. One Pepsi, two *Pepsiat*.

In the first couple of days out of Baghdad, we passed 153 *wireat*. There would be many more to come. The biggest danger came when our driver became complacent and picked up speed. *Wireat* could be almost impossible to spot at a distance in sharp sunlight, and a few times we got it wrong. We rented a fast, low-profile speedboat with a driver called Khalid who had never been south of his village, Aziziyah, and was frustrated at the enforced slowness. When we hit a long straight, he gunned the engine and unknowingly sped under a steel cable at thirty miles an hour. Most of our team were sitting low in the boat, but I was standing close to the bow and threw myself to the ground when I heard Salman shout. The cable caught the brim of my cap and took it to the Tigris. At the stern, behind Khalid, stood a friend of his catching a ride to the next town. The cable hit just above his eye, then sprung on over his head. It knocked him to the floor, briefly unconscious, and took off most of his eyebrow. We dropped him at Zubaydiyah and arranged a car to the hospital. He later called to say he'd got six stitches. If he'd been taller, or Salman hadn't yelled, one of us might have been decapitated. If the wire had been lower it would have flipped the boat and, at that speed, stood a good chance of killing us.

Although the *wireat* were concerning, they were at least a manageable threat. I thought about the different zones of the Tigris we'd seen. Some were geographical and ancient, like passing from the mountains to the *Jazeera*, or crossing the line at Samarra into the floodplain. Others were new, and changeable, like entering areas where remnants of ISIS still operated or leaving behind the checkpoints of the watermelon people. These demarcations were every bit as

real as those geological and historical. We were now travelling in a largely homogenous area, populated by Shia Arabs,
which inevitably meant a reduced necessity for checkpoints
and soldiers. Salman pointed out that if our greatest risk of
beheading came from a wire, then it was at least preferable to
the worst-case concerns we'd had in Ninewa and Salahaddin.

South of Aziziyah, the sky darkened with a film of grey
smog. On the right bank were great smokestacks with red
and white stripes like candy canes, and beneath them were
the scattered concrete and metal of a power plant. Two pipes
flushed water into the Tigris. It was wastewater from the
cooling process, said Salman, and was dumped out a few
degrees warmer than the river temperature. This was one of
the many ways in which the fish and aquatic life were threatened by the plant.

Our visit disturbed a lone elderly security guard who had
been carefully preparing some rice on a hotplate in a dank
concrete room. He made a call and soon a pickup truck
arrived. I presumed the driver was *mukhabarat*. But instead of
asking us to leave, he pulled Ali aside, spoke to him for a few
moments, then got back in his truck. The guard returned to
his rice, and we went back to the boat.

The officer was from a nearby village, said Ali. The
plant covered an enormous area reaching to the outskirts
of Baghdad, and ever since it had opened people had been
getting sick in the villages. 'He thinks there are at least fifty
cancer cases a month,' said Ali, 'but there's no hospital for
cancer here, so people go to Basra. Mostly they're poor, and
they can't afford treatment, but there's no compensation.'

The officer also showed Ali the roof of his pickup. 'It was
yellow and rusted,' said Ali. 'He told me all the metal in the
area is like this. The air is toxic. And the worst part for him
was that finally they sent a company to install a system on

the smokestacks to manage the emissions. That was two years ago, and then recently they had someone come to check it and found out it was fake. He wanted you guys to know, in case you could do something.'

Fishermen punctuated the river every few miles, some with nets and others with electric rods. On the cracked earth of an island, a family of wheat farmers dried their seeds in the sun and tried in vain to get their combine harvester engine to turn over. In the village of Numaniyah, we went to the shrine of the tenth-century poet Abu Tayyib Mutanabbi, for whom the street in Baghdad is named. The story goes that one of his poems insulted a strongman of the day, who eventually tracked down Mutanabbi by the Tigris. He tried to escape, but the strongman shouted after him with some lines from one of his poems, referencing the importance of courage in battle facing one's foes. And so Mutanabbi returned to fight, and was killed.

His shrine comprised six tall columns and a slab of concrete in the shape of a book. It was empty when we arrived, but within a few minutes a bear of a man arrived. He'd heard there were foreigners. He offered us a place to stay, and we followed him back to a gated villa outside town. In the courtyard, a fountain spouted neon-blue washing-up liquid, and another large man sat at a plastic table fondling a shotgun. Two flamingos with clipped wings walked on the grass, and several young deer blinked through the bars of a cage. The house was oversized and looked as if it had been decorated by a wealthy child. A gaudy faux-diamond chandelier hung over a marble staircase, and Chinese terracotta figures stood beside mounted animal heads, Egyptian-themed wall prints and horrific gold sofas. Every surface was filled with memorabilia and vases and silver platters and red London buses and miniature Eiffel Towers, and I thought that I'd never seen so much poor taste in a single room. In that sense, it was a kind of museum.

We didn't know what the man did, and he didn't share. His housekeepers made chicken and rice and we ate it gratefully, then each of us retired into separate rooms while he played billiards with friends as big and garish as the house. It was nice to have air conditioning, but I spent much of the night worrying about whether we were incriminating ourselves with some local mafioso. In the morning we left early and didn't see the owner again. I thought about liberating the flamingos, but I didn't know what we'd do with them, and I was pretty sure a man like this could find us if he wanted to. We closed the gates behind us, and the last thing I saw was the washing-up liquid fountain still gushing figures of eight.

It was just north of the city of Kut when the river became coated with glossy green leaves and small, purple flowers. In places, the plant spread clear across the river like a carpet, only broken by a remembered channel or the movement of a small boat. This was the water hyacinth, often called the Nile flower in Iraq. We had heard of it before the journey. It was a recent arrival to Iraq, introduced less than two decades ago. The water hyacinth floats, absorbs water and forms dense coverage that blocks sunlight and oxygen from reaching the water. In the twenty years it has been in Iraq, it has become a scourge for fishermen and a disaster for biodiversity.

At a bend in the Tigris, we saw a father and son in a small rowing boat. Both wore *keffiyeh* scarves on their heads, and shorts, and the father worked the oars while his son pronged soggy heaps of Nile flower onto the boat with a pitchfork. We idled alongside them, and Ali jumped into their boat.

The man's name was Amar, and this was their major source of income at this time of year.

'The government offered some money for people to do this,' said Amar. He was sweating in the heat, and we moved closer to the bank for the illusion of shade. 'We do three miles on each side. It takes two days,' Amar told Ali, wiping his brow and leaning back in the boat. He was barefoot, and bulbous roots clung between his toes. For those two days of work they'd make one hundred thousand Iraqi dinars, or around sixty-five dollars. They would move downriver as far as they could, but there was competition for other stretches.

The plants clogged irrigation ditches and water pumps, said Amar. They caught in engines and stuck to fishing nets. 'It makes the villages suffer, so they go to the government.' It also grew extremely rapidly. We had read about modified boats that could collect and destroy the plant, but no one we met had ever seen them. Amar said there were people who used chemicals, but that also poisoned fish so wasn't as popular. Occasionally we saw grated barriers placed under water, with just a single opening in the middle for boats. These were useful, Amar said, but really the only solution was for someone to go at it with a pitchfork.

When the boat was full, they rowed to the bank and threw the Nile flower as high as they could into the reeds. Amar agreed that wasn't ideal, because some ended up back in the river. But it was the best they could do. 'You should have seen it fifteen years ago,' he said. 'You couldn't tell there was a river underneath.'

It seemed a pretty rudimentary solution, but it was in keeping with what we'd noticed so far. The river was often managed on a micro-local level. In this case the government had offered money to local fishermen because they didn't have the capacity to solve the problem themselves. The Nile flower was kept in check by people like Amar and his son, in stretches of three miles. Soon we'd be in someone else's territory.

It seemed useful to reimagine all the challenges of the river in small chunks like this. The river had been segmented at every point, by military, farmers and fishermen. To imagine it as a whole seemed both difficult and counter-intuitive. When I mentioned this to Salman he agreed, but said solutions required a recognition of both. Policy should change, and big ideas feed that from the top, he said. But the only way anything actually happens is when local leaders decide to make it happen. *Humat Dijlah* had local coordinators in each city, but eventually imagined their network extending to towns and villages, too.

Just before we entered Kut, a *duba* bridge blocked our path and we got out to walk. A shepherd led us for a while, then we took shade under a wooden lean-to beside an illegal fish farm. 'The funny thing is that we're really against these,' said Salman, pointing to the metal cages under the water. 'They're all meant to have licences and environmental impact assessments, but almost none of them do.' The owner said this farm had 3,500 carp, fed with protein from Holland. But he was uncomfortable, and eventually said he didn't want us to look too closely because of fear of the evil eye. This was a common superstition that any undue attention or compliment could bring bad luck. We told him we'd stop talking about his carp. 'But why did you say it's funny?' I asked Salman, and he giggled. 'Because our job is to stop things like this happening, but if it wasn't for this one, we'd be turning crispy in the sun.'

Chapter Twenty-Seven

Future Youth

Day 51
Kut | Wasit Gate | Gharraf Canal
River miles: 864

The river does a sharp U-turn at the city of Kut, and just before the apex of the bend is the river's southernmost barrage. This regulates volume and divides the water between the main branch of the Tigris and a channel called the Gharraf, as well as smaller irrigation projects. The Tigris then bends towards Iran, while the Gharraf heads due south and joins the Euphrates east of Nasiriyah city. The Gharraf may have been the Tigris's original channel, from thousands of years ago, before its course began to deviate across the floodplain.

Ali reminded me again that the barrage was designed to safeguard against floods. Now it was redundant. It was built in the 1930s, and finished just before the start of the Second World War. Throughout our visit, people would tell us it was built by Sir William Willcocks. In fact, Willcocks didn't actually oversee its construction because he died a year before it began. Also, when he conceptualised the barrage, he was working as the head of irrigation for the Ottoman government. But to many we met he was a Brit, and a good man. There was even a shopping mall in the city named after him.

The British seemed popular in Kut, at least in memory. It was because they left behind infrastructure, said one man. That's why they will always be better than the Americans, he told me, and why fewer Iraqis blamed Britain for the 2003 invasion.* In Kut, there was the site of a major battle between the British and the Ottomans in 1915. The British were heading from Basra to Baghdad and, after a defeat at Ctesiphon, retreated to Kut where they came under siege. As many as twenty-three thousand British and Indian soldiers died, with only eight thousand survivors surrendering after five months of fighting. It was almost another full year before a British force retook the city.

I wanted to ask about the steamboats that came through Kut when it was a port city during the last century, but we couldn't find anyone who knew. There was a museum, but it was closed, and all the antiquities had been moved to Baghdad. So, instead, we sat at a riverside café in the cool of evening. It was thronged with the young and cool, collected in tight groups under neon lights puffing elaborate smoke rings and drinking *Pepsiat*. Now the barrage was a backdrop of soft amber lights, a gloom of lazy water pooled below.

A band arrived in black T-shirts with coiffured hair and gleaming trainers. They were called Future Youth, and they set up with an oud, a violin and an electronic keyboard programmed to heavy synth. From the first note, the atmosphere changed. Idle chatter stopped, and plastic seats scraped closer. The singers alternated, and the band switched from

---

* The British don't escape blame completely. In Baghdad there is a saying: if two fish are fighting in the Tigris, it's probably the fault of the British.

traditional ballads to original electronic music. The prodigy
was the violin player, Karrar. He was lanky, with floppy hair
and glasses with transparent frames, and he played with his
eyes shut, body swaying. Long fingers glided up and down,
never once hitting a bum note.

When they took a break, he came over to see us. Karrar
spoke excellent English, learned from television and books.
When I asked about the books he said, 'At the minute I'm
reading Nietzsche, though I prefer Schopenhauer. I like
Bertrand Russell, too. These are my kinds of books, but it's
hard to get them.'

Karrar was a little guarded with his opinions, but one
of his friends was less so. He was an atheist, he said, and
an artist. Kut was not a nurturing environment for those
choices. 'When I say there are no opportunities, I mean absolutely nothing,' said the friend. He smiled. His accent, like
Karrar's, was tinged with colloquial Americanisms learned
from Netflix. Everyone in Kut was conservative, he said. No
one thought for themselves. His parents supported Sadr, and
so did every other person they knew. What did he think of
Sadr, asked Emily, knowing the answer. 'That he's a fucking
moron,' said Karrar's friend.

Most of the audience were in their mid-twenties, and, like
Karrar, probably also a little glum about their lot in life. A
few were harder to peg. There were two older couples, who
sat quietly tapping their feet, and a father and son dressed in
matching double denim. The boy, I noticed, had a printed
picture of his own face on the back of his mobile phone
cover. They sat at the front.

That night was the happiest I ever saw Salman. He sang
and danced, and laughed until he was red in the face. Hana,
too, seemed to lose herself in it. Emily and I didn't know the
words, but she joined in gamely. She also seemed happy here.

Close to midnight, Karrar asked: 'Do you want to go some-where cool tomorrow? It's my favourite place in the world to play music. If you have time, let's take a trip.'

The morning broke early and hot as we drove south-east out of Kut. The heat intensified away from the river and, as miles passed, the lush green wilted then died. Soon, we were in a desert. Karrar was taking us to the eighth-century city of Wasit. In Arabic, Wasit means 'middle', and the set-tlement was built halfway between the important cities of Basra and Kufa.

We arrived to dry mounds and craters that rolled out into the haze. In the middle, breaking arid monotony, was a single, tall, ornate gateway. Once it had been the entrance to something rather grand. Two engraved columns of mud-brick flanked a series of recessed arches, each decreasing in size until they opened to a humble doorway at the bottom. Now it led only from one part of the desert to the next.

Karrar wore oversized sunglasses like a rock star on holiday and swung his violin case as he walked. Around the structure were low walls marking its past life. Karrar called it Wasit Gate, but really it had been the entrance to a religious school called Sharabei that was inside Wasit city. Some parts of the gate had been unearthed from a tell in 1964, and enthusiastic state archaeologists made their best guess at piecing it back together. That explained why it was a little wonky, and why the right tower had exquisite pat-terns of diamonds and the left did not. On closer inspection, we found modern yellow brickwork used in the reconstruc-tion, and crudely cemented in.

There was an Ozymandian power to it, standing alone in the scrub. Karrar led us around the back, where inside the more authentic of the two towers a staircase wound up

to open sky. From there, the blueprint of Wasit was clear,
stretching for miles. 'I love it up here,' said Karrar. 'It's my
favourite place in the world.' He came with friends, he
said, and sat atop the gate and imagined the ancient city. I
wondered if to him the spectre of the place long gone was
exciting because of the disappointments of his contemporary city.

He took the violin from its case and stood on the very
precipice of the tower. 'I wrote a song about this place when
I was nineteen. It's called "Nostalgia".' He began to play,
and swallows nesting in the arch swept out to join us. The
song was slow, and sad, with long, vibrato notes in a minor
key. I looked from Karrar to the outline of the old school
below, and then across the expanse to where morning sun
bleached out the horizon. I tried to imagine the city alive,
as Karrar did. Ibn Battuta had been here and wrote: 'It has
fine quarters and an abundance of orchards and fruit trees,
and is famed for its notable men, the living teachers among
whom furnish lessons for meditation.' It was hard to visual-
ise anything living out here now.

Wasit was, like all Mesopotamian cities of antiquity, built
by a river. The Tigris once flowed through and encouraged
construction of the schools and palaces of the Umayyads
and Abbasids. The city was an important military and
commercial node in the networks of the caliphates, and
became a hub for boatbuilding. And then, in the fifteenth
century, the river pulled away. Within a century, it was no
longer sustainable for a city to survive here without the
water. The fields and channels were left to silt up, then to
dry completely. The inhabitants left, and wind, sun and
time did the rest.

If ever we needed a reminder of the life-giving power of
the Tigris, and what happened if it was taken away, here

it was. History left these lessons imprinted on the earth, marked in the soil like cuneiform, and it was up to us to read clearly and take heed. No one in Wasit, nor the Caliph who built it, would have imagined that one day it would be buried under cracked earth. Few in Basra or Baghdad really believe now that the same could come to them one day. Ali had always spoken about the wave of refugees that would follow an ecological disaster on the river. 'If the river dies,' he'd told me earlier, 'Iraqi people will be forced to leave, and that's an international crisis. If we could bring stability into our country, and avoid more conflict, we'll ensure that people will stay here.'

We ate lunch at the home of a sheikh who lived in a village atop the ancient Tigris. He had invited us after hearing we were at Wasit Gate, and we squeezed into a long, narrow eating room with about seventy other men. Two lambs had been killed, and the heads and other delicacies were brought to us. I wondered why the most desirable cuts of meat here always had to wobble so much. The sheikh, Jaafar, was so keen that I try it all that he fed me, pressing a gizzard into a ball of rice and prodding his fingers towards my mouth.

The villages under the sheikh had a population of three thousand, and they grew the usual wheat and watermelon and herbs, and kept sheep and goats. Their water came from the Tigris through a distributary and then a canal, which was always drying out, he said. They used the traditional method of flood irrigation, in which the channels are constructed in such a way as to raise the level of the water which can then be taken into fields by gravity alone. It was managed by a rationing system which during lean months dictated when and how much each village could use. Occasionally, lack of water led to disputes. In 1998 and 2017, there had been serious family feuds in the areas that led to killings. Not that far away there was a bigger conflict that had led to twenty-three deaths and opposing villages firing rockets at one another.* It was unresolved. In 2018, most of Jaafar's animals died because of drought. This year there was just a trickle left, he said, and he was going to have to invest in buying water from elsewhere.

Jaafar's irrigation method was outdated, but he was not alone. Many farmers that we had met were using flood irrigation, either as Jaafar did, or now more commonly via pumps that pulled water from the canals or river itself. Jaafar was proud to use the same methods of irrigation that had been successful in Mesopotamia for millennia, from the time of the Sumerians, but now he had pause for thought. It required so much work to keep the canals and drainage clear, and in recent years the saline levels in the water and on his land had become unmanageably high. Something had to change, he said.

Irrigation is responsible for almost 80 per cent of water usage in Iraq, and for years experts have been encouraging alternatives to the highly wasteful flooding system. One solution would be the widespread introduction of spray and drip irrigation, which uses just a fraction of the water that flooding does. I asked Jaafar about this, but he dismissed it quickly. It required enormous investment, he said, and the government should be responsible for that. But they had no interest, or no money, he said, and he couldn't afford it himself.

His farm was rowdy. Outside, a couple of cars turned up with young men carrying guns. They'd been hunting birds, they said. They fired their Kalashnikovs in the air and, when they saw that we didn't really like it, they did it again. Somehow the desert made it feel more lawless and dangerous

---

* It's also true that often these disputes are between families or tribes with pre-existing bad blood, and are not always purely because of water rationing. But the shortages bring feuds to the surface, and have been responsible for them turning deadly.

than the life we knew on the river. I really wanted to use the toilet, but the only available outhouse had thigh-high brick walls around the long-drop, and nothing else. I decided to wait rather than have an armed audience.

Jaafar drove us away from his farmland and parked beside a tell. There were at least ten thousand of these tells in Iraq. This area had been most heavily populated during the Abbasid caliphate. The ancient river and its branches were like a spider web, he said. 'It's full of treasures,' he continued, leading us against a robust wind. Under his sandalled feet were shards of green pottery mixed with firm, dry soil. I picked up a piece. 'It's an Abbasid jug part,' he said dismissively. 'If you stay here an hour, you'll find a complete one.' One man he knew had found gold coins, and promptly sold them and moved to Baghdad. There were also fossilised creatures from the old riverbed, he told Salman. 'Thousands of people lived here,' said Jaafar. 'They had everything, but the thing they needed was the Tigris. That's all we need, too, or we'll end up like this, and in a hundred years someone can pick over my pots and pans in the dirt, too.'

We drove west to a town called Al-Hay, from where we could follow the Gharraf Canal back to Kut and then continue along the Tigris. Karrar knew of a small museum in town, and while we waited for a fisherman to prepare his boat, we knocked on the door of an elderly *haji*. His museum consisted of a single room, piled high with handicrafts, old furniture and framed pictures. He wanted to remember everything that had once made Al-Hay special, he said.

'What's been lost?' asked Emily.

'People,' he said bluntly. There had once been a large Jewish population in Al-Hay. By 1947 they had all gone, driven out, but the *haji* remembered them well and fondly.

There was a synagogue in town, and several houses owned by families who now lived in Israel. Those homes were permanently locked up, and the paint was peeling, but nobody broke in or stole anything, he assured us. It was as if the owners hoped they could return, and the residents of Al-Hay did, too, so everyone agreed to watch over the buildings. The *haji* showed us an old wardrobe that he said was Jewish, and a few other items of indistinguishable age or provenance. It seemed little to remember a people by. Without the wardrobe, though, perhaps there'd have been nothing at all.

His son brought us tea, and I sat on a rocking chair. They were talking about Fayli Kurds now, Shia Kurds who lived predominately on the border with Iran. There used to be Faylis in Hay but, like those on the border, they were deported and killed in large numbers by Saddam Hussein. At times, the bloodshed in Iraq, in the name of ethnicity or religion or politics, felt impossibly heavy. All that was left from the entire history of two communities was now contained in this dusty room, sharing shelves with baskets from Nasiriyah and pieces of Abbasid pottery.

Iraq's history showed two distinct patterns. One was of the coming together of different peoples from north, west and east; from the earliest Mesopotamian civilisations to the Persian, Turkish and Arab migrations of more recent centuries. The second was how co-existence had turned to conflict, and a persecution of those deemed different. The country still had Kurds and Arabs and Assyrians, different Muslim sects, Christian denominations, Yazidis and Zoroastrians, and other minorities in the north. But so many were suffering. And how should the Jewish community be remembered in Iraq? Or the dwindling numbers of Faylis left? And how could there ever be justice for what was done to them? Few people we'd met in Ninewa and Salahaddin offered leniency to prisoners who had been sympathetic to ISIS, regardless of circumstances. Justice had been so rarely delivered in Iraq's recent history, so it followed that forgiveness might also be slow to follow.

Iraq has a modern history of oppressing minorities, sometimes until they are completely wiped out. But Salman encouraged me to also remember how hospitality was the opposite of persecution. It was the ubiquity of hospitality that had first brought me to the region, then kept me there. When I looked for something hopeful in the country, I found it in that kindness that was still present in homes all across Iraq and which superseded religion, ethnicity and anything else in its way. How could that both of these things be true – the violence and the kindness? This was part of the complexity of Iraq, and I may never understand it.

We returned to Kut by boat along the Gharraf under a burned sky. There were unfamiliar sounds going upstream. The water slapped the hull more aggressively. Wind came at us from a different side. Black and white kingfishers hovered when we stopped, wings beating so fast as to almost disappear, leaving the body suspended. In Kut, Karrar went home, and Ali left us, too. He had work to do with *Humat Dijlah*, because there was only so long both he and Salman could be gone. Ali wished us luck, and we were back to five.

Chapter Twenty-Eight

Electro-Fishing

Days 52–55
Sheikh Sa'ad | Ali Al Gharbi | Ali Al Sharqi
River miles: 968

Isam was the first man I could remember seeing wearing shorts. They were bright, Hawaiian, and his T-shirt bore the name of an American netball camp. Around his neck hung an eagle talon on a chain, and on his waist a leather belt and pistol. Isam could get away with this unconventional look because he was a policeman, working the river around Kut. He would be with us all the way to the city of Amara, four days downstream.

He had been with the river police for seventeen years. The Americans taught him how to dive, in the Tigris, and he spent three months in the Green Zone in Baghdad and trained in Jordan. Now he lived on a house by the river and swam every day. He knew this stretch better than anyone, above and below the surface, and was excited for us to see his Tigris.

Modern houses crept close to the edge of the riverbank. Farther back were traditional mud-brick buildings, squat and long, now occupied by cattle and goats. There were fishermen out, too. Some used nets, which they flung out across the broad course of the river, and Isam slowed each time to carve around the small buoys that marked territory.

Wheat season was over, and farmers were at different stages of switching crops.* Some were burning the fields. We passed walls of fire and smoke, with young men skirting the edges wafting flames in the right direction. Others were a day ahead, using pipes to soak blackened earth and crouching to mix chemicals into the soil. In a few days the watermelon seeds would go in, and in three months the fruit would be ready.

Nestled in the reeds close to a village called Sheikh Sa'ad was a red *guffa*. It was made from iron, and its owner a jolly man called Abu Jassem, who quickly paddled over to see us. 'I'm like a fish,' he said, by way of a greeting. 'I'm always in the water.'

He had made this *guffa* himself. When he was a child, his father assisted a specialist in the traditional method. 'He'd use pomegranate branches to brace it and be strong,' Abu Jassem said. Inside and out were sealed with black tar after the reed hull was woven. Leaves from date palms lined the interior. Like everyone else, he hadn't seen a *guffa* since the early 1980s. 'Before you were born!' he said, looking at me. That was just about true.

He rolled this *guffa* out into the middle and paddled vigorously on one side, so it spun like a top. His feet were planted on the concave base, knees bent, and he worked with gravity to let the boat twirl around him. I got in alongside and immediately stumbled. Any balance I thought I had was deserting me. Abu Jassem loved the uncertainty and set off frantically across the river, short arms making muscled strokes on alternate sides. His style and power defied the

---

* In hindsight, the farmers that we saw were the lucky ones. In 2021, almost 40 per cent of wheat farmers in Iraq experienced an almost total crop failure.

current. Then, some distance from our boat, I was tasked with bringing us back. Instead I spun us around uselessly, drifted downstream and nearly fell out, and the paddle was taken away from me.

'You're not an Iraqi yet,' said Abu Jassem when he had guided us to the police boat. He was smiling. 'Using a *guffa* is important. All our life depended on this once. My father used this to fish and to feed us. For me it's still important for crossing the river.'

He was still beaming and spinning until we rounded the next bend. Despite my poor showing, I was delighted to have finally seen a *guffa*.

Isam took his job seriously. When he spotted a fisherman that he suspected of using electricity, he gave chase. 'Give 'em hell!' shouted Emily, and Isam accelerated into the middle of the river, fast enough for the bow to rise and the keel to bounce across the water. It didn't take long to reach the boat he'd spotted, which was now partly hidden in the reeds. Two men, rather red and breathless, stood awkwardly on the bank.

Isam knew them, he told us. They were trouble. Their boat was empty and they denied fishing with generators, though Isam was sure they'd hidden their kit in the reeds. In the cool box were a dozen large fish which would not have been possible to catch with a simple net. 'I've been arresting these bastards every few months for six years,' Isam sighed. They lay back in the boat and said nothing. Both had been in prison, for these crimes and others. 'I see you more than my family,' he told them, and one man laughed. But there was nothing more to do, unless we wanted to tramp through swampy reed beds. 'I'll be back,' said Isam, and the criminals clicked their tongues at him and turned away.

In Sheikh Sa'ad, Isam picked up a colleague called Uday.

Occasionally the electro-fishermen got violent, so he wanted backup. This was the most time-consuming part of his job. Sometimes he confiscated equipment and, if it seemed warranted, perpetrators spent a day or two in prison. Once, when he was closing in on a boat, the fisherman panicked and jumped overboard, straight into Isam's motor. 'His hand came clean off,' said Isam. Uday nodded and made a gesture to show how a hand could be removed by a motor. Isam's tribe had to pay blood money, even though he was acting as law enforcement. 'Our tribal system is more powerful than those rules, and I didn't want a feud. I still see the man, and I wave to him every time now. The bastard is still out, illegally fishing, but at least he's slower now.' Isam winked. All his stories were told with a wry smile, as if no matter how bad or sad his job could be, there was always a chance to laugh, call someone a bastard and light another cigarette.

Isam caught six electro-fishermen during his time with us, and we all agreed that the chases were the most fun we'd had in weeks. He stood at the wheel, head always tilted up so he could look long and straight down the river, and then with a sudden movement he'd ram the throttle. There was a thrill because boats could disappear into reeds, or down arterial channels, and there were still *wireat* and fishing nets regularly crossing our path. But Isam loved the challenge.

I was surprised that often the electro-fishermen we caught were meek and quiet. Uday pulled generators or batteries out of their boats, and we hauled them onto ours. Wires were ripped out of oar pins, and the charged net and pole slid onto our deck. Finally, any paralysed fish in their cool boxes were thrown back, and this was the only time there was any real complaint. Couldn't you just let us keep the ones we already have, they pleaded?

'You know this is *haram*?' asked Salman of a father and

son in torn clothes. The old man shrugged. 'We don't have another option,' he said. 'Do you want us to eat the air?'

It was easy to feel sorry for them, because they were poor and only subsisting from what they caught. But the river was also busy with other fishermen toiling with nets. They had the same reality of diminishing fish stock and low water levels. And yet they used traditional methods, requiring skill and craft. Those using electricity were not just finding a way for their own families to survive, but also depriving others. It was as selfish as it was desperate.

The gear was expensive. Generators were roughly four hundred dollars, batteries over six hundred. Some would have to be returned to the fishermen if the tribes got involved but the rest were kept in the station. One evening near the border with Maysan Governorate we chased a boat for three or four miles before it slalomed into reeds. The driver must have called a friend to bring a pickup truck to take the equipment. But we arrived too soon, and the truck was still empty, though so was the boat. Our team was deputised into an active role and fanned out. Isam returned with the wired net, and soon after the generator was found behind a bush. The only missing piece was the wooden convertor box and I burrowed deep into the reeds, keen for glory, trying to ignore the spiders tumbling down my shirt. When I saw a red wire I hollered with joy and burst back out triumphantly with the proof. The man shrugged, silent now, and Isam let him go with a warning. Back in our boat we celebrated with juice and snacks at another successful bust.

At night we slept in the homes of Isam and Uday's friends. It was still Ramadan, so we shared typically abundant *iftar* meals and often entertained large groups of curious villagers who came to watch us eat. Emily and Hana would usually retire to the women's side, where they were given space to rest and relax. In Ali Al Gharbi, they spent a few hours sitting cross-legged in a line with the daughters of a family plaiting each other's hair, then everyone went to bed early. Meanwhile, Claudio, Salman and I sat in a room with eleven loud men talking about politics and religion. Our host, Mohammed, was enormous and excitable with a terrifying voice and a lot of opinions. I was asked about religion in Northern Ireland, how many wives I could have, or wanted, and what I thought of Sadr. How would Sadr do in British politics? Did we have anyone like him? I said we hadn't, though I wanted to say we had plenty of other buffoons and charlatans. I tempered my answers to be bland hoping I'd be left alone, but they kept coming. Did I know the Americans had brought the Nile flower to suffocate Iraqis? How about the fact that Israel funded ISIS? And when on earth were Manchester United going to rekindle their form?

At 2 a.m. I got into my sleeping bag and at some point the other men finally lay under their blankets, too. The rest of the night was a cacophony of snoring, snorting, farting and the tinny sound of Facebook videos playing on smart-phones. It always surprised me how little men seemed to sleep in these villages. This pattern of staying awake late, scrolling on phones or calling other insomniac friends during the night had been common throughout the country. My alarm woke me each day at 5.30 a.m. I was often exhausted come morning.

There was also never any personal space. An Iraqi friend of mine said this resulted from the tribal system, and that space and quiet for an individual was not highly valued. It was much more important to be surrounded by a commu-nity of others at all times, and to feel comfortable around one another. As an only child from a tiny family in rural Northern Ireland, this took some getting used to. It was now

45°C in the heat of the day, and increasingly difficult not to drift off on the boat after these disturbed nights.

As we approached Amara, Isam pointed to the base of the reeds. There was a fresh yellow mark, he said, two or three inches long. That was how much the water had dropped in the days since we left Kut. He was an observant companion and pointed out the first and only wild boar that we saw, its enormous, coarse rump blundering into a high reed bed as we drew level. The reeds were taller than normal, said Salman, because of the pollution, which encouraged unnatural growth in reeds and algae, and choked the ecosystem of the river. Isam said he didn't know about that, but there was plenty under the water that shouldn't be. He'd pulled a motorbike out of here once, and a car not far away.

Uday told us he had a small boat of his own, which he called *Dijla*. Unfortunately, he said, it hadn't left Kut. Many people we met along the river had never seen the Tigris beyond their own area of influence. Perhaps they had been to Baghdad once, or driven a road that ran alongside, but even this was rare. For a lot of the pastoralists and fishermen that we spoke to, their lives were centred around an area of a few square miles.

Those same people sometimes spoke of Turkish and Iranian dams, which must have seemed so abstract. They blamed the poor water situation on projects in mountains they had never seen. It was perhaps no wonder that many treated their own section of the river in isolation; here, in the heat and the sparse flatness, the idea that this same river once carved through snow-capped limestone mountains and forests of oak must be almost impossible to imagine.

Isam made fun of Uday for his boat called *Tigris* that only travelled only two or three miles. I told them I knew of two other boats with the same name that had somewhat grander ambitions. One was a twenty-six-feet-long reed boat built by the Norwegian explorer Thor Heyerdahl in 1977, which he sailed from the Tigris–Euphrates confluence to the Persian Gulf and Arabian Sea. His ambition was to prove that the ancient civilisations of Mesopotamia, Egypt and the Indus Valley were in contact with one another, and that reed boats crafted in southern Iraq could sail the open seas.

In 143 days, and over four thousand miles, Heyerdahl and his crew were successful in reaching the Horn of Africa, by way of the Indus Valley. At the port in Djibouti the reed ship called *Tigris* was burned, much as in the Viking funeral myth, as Heyerdahl's protest against modern warfare. Rashad Salim, whose knowledge of traditional boatbuilding in Iraq had allowed us to trace the journeys of the *keleks* and *guffas* along the river, had been crew on that expedition, and it was perhaps that experience that encouraged him in the importance of protecting Iraq's maritime heritage.

Uday said he'd never burn his boat, and Isam said that foreigners sometimes had very crazy ideas. The second boat called *Tigris*, I told them, was also part of a grand vision. In 1836, a British army captain called Francis Rawdon Chesney was chosen to scout new routes through the Middle East to India. He had two huge paddle steamers built, one *Tigris* and the other *Euphrates*, then dismantled and transported overland from the mouth of the Orontes River in Syria to a navigable stretch of the Euphrates. It was a distance of 140 miles, through desert, marsh and mountain, and much like Herzog's *Fitzcarraldo*, seemed a mad undertaking.

Improbably, they made it, and began an almost equally daring journey down the treacherous and relatively unknown Euphrates. On 21 May 1836, the two steamboats were caught in a freak tornado, and the *Tigris* became stuck in the current where it succumbed and sank. Twenty men died – though

Chesney somehow survived – and the *Euphrates* steamer continued alone to the Gulf, whereupon the British government scrapped Chesney's plan to use the river. Later, steamers did begin to use both rivers, which some see as vindication of Chesney's efforts. Mounted in the British Embassy in Baghdad is a piece of broken marble that was found in Basra in the 1960s, from the *Tigris*, which has engraved on it a commemoration of the event. Aside from that, the story is not widely known.

Isam had listened carefully, and his thoughts hadn't changed. Foreigners had some crazy ideas. But he did think Chesney's trip had more value. He was trying to achieve something, Isam argued. I said I felt like it showed the folly of imperial ambition. Heroic and brave, certainly, but ultimately colonial hubris. He shrugged. 'We always need to advance, and look forward,' he said. For him Heyerdahl's trip was meaningless. There was nothing to learn from the experiment. We agreed to disagree, and I asked Uday what he thought. 'I think I'll change the name of my boat,' he said dryly.

Chapter Twenty-Nine

Baptism

Days 56–58
Amara
River miles: 1,010

We arrived in Amara on *Eid Al-Fitr*, the end of Ramadan, and went to mosque in the city centre. The sun rose heavy like a hot-air balloon over the Tigris, bringing with it stifling heat. The mosque was old, but the city had grown around it so tightly that there was no way to get a perspective of size or style. Its entrance was adorned with green flags and posters, and by the doorway sat a dozen women in black abayas. These were widows of wars and sanctions, from Iran to ISIS. They waited in 45°C on the morning of Eid for charitable donations from the faithful.

There were a few soldiers, too, and a man wearing a green headscarf. This suggested he was a *sayeed*, descended from the Prophet Muhammad, and was not unusual. There were thousands of *sayeeds* around. He told us that coronavirus was sent by God to balance out the world and make it easier for the Mahdi to return. There were five more great disasters to come, so now might be a good time to join Islam. I said I'd think about it. We were the only people I could see wearing masks, and I thanked the *sayeed* for his wisdom and went inside.

The case numbers of Covid-19 seemed to have dropped in Iraq, although there were so few tests carried out that it was hard to tell. Batches of vaccine had arrived but uptake had been slow. We'd heard there were thousands of doses of Astra Zeneca in Amara but only a few hundred had been administered. The population was largely sceptical. That was likely to change, however, because Sadr had recently taken his dose live on TV and encouraged his millions of followers to do the same. Iraqi social media soon became swamped with video of Sadrists holding his photograph while getting the jab. My favourite was a man who had a tattoo of Sadr on his shoulder and received his vaccine shot through Sadr's nose.

A hundred men gathered inside the prayer hall. Women were not allowed, though an exception was made for Emily and Hana. Ramadan had been broken at *iftar* the night before, and now there was a sermon to mark the celebration of Eid. Most men stayed only for the prayers and then shuffled out into the morning light after fulfilling a minimum requirement. We went to speak to the imam after his sermon, which had lasted well over an hour and perhaps explained why so few people stayed.

He was a sombre young man, and he stood heavily under a large wall print of Quranic script. He sighed a little when we arrived. Mercy and peace were important at Eid, he told us, and as Christians we were all brothers under Abraham. When Salman translated this to me, he said, 'The Imam welcomes you as brothers, because Christians and Jews are the brothers of Muslims under Abraham.' The imam stopped him. He wanted it to be clear that he had not included the Jews.

This was rare. Throughout Iraq, I was regularly told of the closeness of the Abrahamic faiths. Though Iraq had eradicated its Jewish population, I often heard people express regret. No one spoke of the pogrom in Baghdad, of course, the pro-Nazi *farhud* in 1941 during which two hundred Jews were killed, nor of the increasing persecution the community had faced in the lead-up to the creation of the state of Israel. The Iraqi Jewish population, whose history goes back to the sixth century BC in Babylonia and whose number was at least 150,000 in 1948, almost completely disappeared over the course of a couple of years. Most emigrated to Israel, where there is still a large community. People rarely mentioned this. But I had been surprised at how often the Jewish heritage of Iraq was conjured by those we met along the river, and cited as one of the tragic and irreplaceable losses of the country's history.

Now the imam said that because of Israeli bombardment of Gaza, which was ongoing, he could not include Jews in the brotherhood of peace. Salman pointed out that the actions of the Israeli state and the Jewish people should not be conflated, but the imam was clear: Jews were not included. Later, he called Salman to say that he'd been reflecting. There were other people in the mosque listening to our conversation, he said, and everyone cared about Palestine. So he felt the need to make a firm statement. But now it was just us, on the phone, maybe he could be softer. He was tired from all the prayers that he had to make for Eid. We elected not to return for another conversation.

Every Sunday in Amara there is a baptism in the Tigris. Those who are renewed in the waters are the Sabean-Mandaeans: the smallest ethno-religious group Iraq. The population of the Mandaeans is tied to their sacred life source, the twin rivers, and they have historically lived on the Tigris in Maysan and the Euphrates in Dhi Qar, though there are pockets strung out along the river as far south as Basra.

In the south of Amara, on the east bank at dawn, around thirty Mandaeans gathered by the water. The men wore loose white linen shirts and trousers, bound by rope, and they padded around in bare feet. Their heads were swathed in white scarves and the women dressed much the same, through their clothes were even more voluminous and flowing. There was a space for everyone to sit on concrete benches under an iron roof, but most stood by the gentle steps leading to the Tigris. Across the river the city slowly awoke, sounds of car horns and loudhailers and muezzins drifting back, all unaware or unbothered by this ancient ritual taking place.

The priest, whom the Mandaeans called sheikh, was Bashir, and his long beard was hidden by the scarf around his chin. In his left hand he held open a small, string-bound book which he read inwardly. His assistant, also with a beard that reached his chest, lit a small fire around which they both stood. Bashir said he was happy to speak with us, until the point at which the baptism started. After that, he would be occupied.

'We are the oldest religion in the world,' he said. This was at least the third time that a religious authority figure in the Middle East had told me this, but I took it in the spirit in which it was intended: we are very old, and predate the Abrahamic religions. There were baptisms, *masbuta*, every Sunday, said Bashir, for any members of the community who felt they needed it. Most Mandaeans only felt comfortable going six to eight weeks between immersions. Some here were from Amara, and some had come from Basra.

'Our religion comes from Adam,' Bashir continued. They believed God gave a message to the angel Gabriel which he delivered to Adam, and his teachings from the Garden of Eden were passed to his son and eventually down to today's Mandaeans. Abraham was condemned for being willing to kill his own son, and their prophet is John the Baptist, revered much more highly than Jesus. 'For us, baptism is cleansing, and it should happen regularly,' said Bashir. He turned his body to the dancing flames and recited something under his breath. The assistant indicated they were ready to begin.

First, they used a brush to sweep the steps by the water so no one slipped which, although not part of the sacred practice, seemed sensible. The women were to be baptised first and lined up, eight adults and two children, then followed Bashir tentatively into the water. One by one he held and then gently dunked them into the Tigris. There was no drama to it, just simple immersion with a few whispered words. As they lifted their heads out, most took a mouthful of water, and I couldn't help but think of everything we'd seen going into this river.

The women were held a while by Bashir, who said to them: 'Listen and request, and you shall find it.' Then, 'The truth will heal and strengthen you.' When each woman first left the water, she came to the fire and walked around it three times, forbidden from getting too close. Incense sticks burned, and the scent drifted towards me as I watched. Each rotation had a meaning: first, for peace and light, second a blessing for people and their households, and finally a prayer to the angels. Behind, the city seemed to have fallen quiet, and the sound of the bodies in the water and the lapping of the disturbed river water was all I could hear.

When all the women had been baptised, they sat together on the bench, and Sheikh Bashir walked along their line. He carried a wooden stick, which he called a *margena*, and the green myrtle leaves. 'Every part has its secret,' he'd told me earlier. His clothes, or *kisa*. Something that looked like an olive-wood cross: *darfush*. A piece of silk cloth; all had meaning. The string-bound book held the key, but it, too, of course, was only for Mandaeans.

Now the priest whispered in the ear of each woman, using a religious name unique to each one, created through a spiritual calculation based on the exact time and alignment of the stars on their birth. The name should not be known by anyone other than the priest and family, though one of the young girls kept saying hers out loud because she liked the sound. Myrtle leaves were rubbed on foreheads and tucked under headscarves. The women went back to the river and washed their right arms from elbow to fingertip, then held them straight out perpendicularly. A gold dish was passed around with river water, from which a drop was drunk and another poured over the left shoulder. Finally, Bashir prayed over them with the holy book, his right hand resting on each head while they continued to hold arms aloft. Then it was over. The priest returned to the fire, and the assistant went to check on the food his wife was preparing. The sound of shuffling behind us suggested men were preparing for their turn.

For the Mandaeans, water is divine, and the Tigris the essence of their faith. Temples, *mandis*, are built by the water. Theirs is a dualistic world of light and darkness and the soul lives in exile, trapped inside a human body, only escaping in death. The rituals around death are key to sending the soul safely to the next world. There are several holy texts, of which the *Ginza Rabba* is the most important.

This is what I gathered from my snatched conversations with Bashir and with another Mandaean called Farhan who sometimes performed the role of the assistant. He wore fake Ray-Bans, and his beard drifted in the light breeze like an inverted poplar. 'The water here is the same as the water in the next universe,' said Farhan. 'Life couldn't start without this water. Rivers and the sea give life.'

This foundational element of faith has been challenged in recent decades. There had long been an established Mandaean presence in Baghdad, where they were tradition-ally associated with silversmithing and boatbuilding. They were respected, said Farhan, and had no more problems than anyone else. I'd read that Saddam protected them because he employed a Mandaean to cast spells for him, but here the community were adamant there was no magical element to their religion. It reminded me of the Yazidis who had been branded devil-worshippers out of a lack of understanding of their faith. Mystery in religion has so often been assumed to be dark or evil, then punished out of ignorance.

Things got worse after 2003, said Farhan. Mandaeans were targeted by both Sunni and Shia militia groups when the sectarian conflict began. They endured kidnapping and robberies because they were deemed to be wealthy, and the tolerance of their minority diminished. From a population of thirty to fifty thousand in 1990, the best current esti-mate of their number within Iraq is around five thousand. There are sixty thousand Sabean-Mandaeans living outside of Iraq, in at least twenty-two countries. That diaspora has had to recreate the elements required to sustain their faith in strange environments. Baptism must occur in running water, so communities tend to coalesce around rivers in their new countries. But in riverless Erbil, where there were a few hundred Mandaeans who had fled Baghdad, they resorted to running a pipe in a circuit through a modern concrete temple to create the illusion.

The mass migration put pressure on priesthood, too. Years of training are required, and many priests have left to be with their communities in the US, Australia, the UK and else-where. The high priest of the Mandaeans, Sheikh Sattar, now lives in Australia. In Amara there were six priests left, but in

Basra all had gone. That explained why the Mandaeans from Basra were travelling two hours north for baptism.

I sat with a woman called Majdi, whose priest had fled to Jordan a year earlier. Her skin glowed after the immersion. 'To be honest, I don't mind,' she said of the travel. 'The water here is more pure. In Basra it's so salty.' I asked if she worried about being poisoned by the river, or getting skin irritations, but she laughed it off. 'If you're pure, you'll come to no harm in the water.' To her, baptism was new life each time. It washed her soul and she came out clean. She was with her six-year-old daughter, who she hoped would continue to follow the faith. Only men could marry outside of the community and keep their religion, so her daughter's options later in life would be limited, but again she wasn't worried.

'The thing is, life is good for us. In Basra, they treat us well. There are good opportunities and jobs, and no one discriminates.' Mandaeans have strict religious rites, but live a similar everyday life to other Iraqis. They have phones and cars and jobs and hobbies, said Majdi. They were Iraqi, and human, and that mattered most of all when it came to co-existence. Here Farhan, the erstwhile assistant, cut in. 'But Mandaeans are leaving. And we need help. We need schools and services and support. But all of Iraq is going through this. We hope it'll get better.'

He mentioned the language, Mandaic, which is a dialect of Aramaic unique to the Mandaeans. 'This is really what we need to protect,' he said. 'All of our secrets are related to the language.' Some juice was brought, and we sat together in the shade watching Emily, knee-deep in the Tigris, photographing the men's baptism. Farhan had some ideas and hoped there could be funding for language schools for Mandaean youth. There was a brief silence, then Farhan said he was a Liverpool fan, and the only thing that would encourage him to leave Iraq would be if he could move to a house beside Anfield.

In a covered hut, fish washed in the river was cooked in silt from the banks. This preparation was also part of the ritual. Emily was called from the water to try it. Later she said the silt made the fish taste like faeces, which was maybe unsurprising given what we'd seen go into the river. It was one of the worst things she'd ever eaten. She tried to rush down the first serving quickly, swallowing each piece in one go to avoid the taste as much as possible. But this was a rookie error, and she was given more as a reward for the haste. She has not had *masgouf* since.

Chapter Thirty

The Marsh Arabs

Days 59–61
Qa'let Saleh | Al Cheka | Hawizeh Marshes
River miles: 1,050

Our boat left Amara in semi-darkness and the city slipped away with ease, taken by the reeds. As light flooded the day, ducks and geese joined us in the current. Occasionally we passed factories making bricks, their chimneys exhaling black smoke that diffused the sky. This was about the worst job in Iraq, said Salman. It was hot, hard, noxious and badly paid. Every time we went past one, I thought of the poor souls inside with no other options.

The Tigris gradually shrank as channels branched off to feed the Hawizeh Marshes to the south-east. Maysan was home to the majority of the eastern Mesopotamian marshes, as well as being an important governorate for oil production. In the town of Qa'let Saleh, we stopped by a broken lock. An old shepherd said that our modest, small-engine speedboat was the largest he'd seen on the river since the 1980s. 'Nobody uses it any more,' he said and moved on.

We squeezed through under the broken sluice gate, and on the other side disturbed a family dunking their sheep in the water to keep them cool. The animals, sweltering in their coats, panicked and struggled as they were carried to the water, then immediately relaxed in the cool flow. Most of them celebrated by defecating, and we floated through into a pond of shit, piss, bedraggled sheep and confused shepherd boys.

Officially our boat was privately owned, but the captain had been suggested to us by the *mukhabarat*, and as with the other undercover agents he couldn't help but look like an intelligence officer doing an impression of a civilian. He was friendly, though, and didn't interfere, and was talented when it came to avoiding sheep and *wireat*.

We arrived at the edge of the marshes at night and stayed in a village called Al Cheka. With a thick white beard and pristine chequered headscarf that shrouded his eyes, our host Abu Hasan spoke with youthful expression, though heavy, stiff hands betrayed his seven long decades. For as long as he could remember, he said, the waters of the expansive marshes of southern Iraq and Iran had been receding, held back by dams and berms, the rivers that fed them diverted upstream. But after the fall of Saddam Hussein's regime in the spring of 2003, it seemed for a while as if they might be coming back.

'When we heard he had gone, there were around twenty of us that started,' said Abu Hasan, sitting cross-legged against the back wall of his guest room. 'There were no shovels, so we had to send someone to borrow them from the Ministry of Water Resources. Then we stood on the dam and dug and dug and released all the water back.' He smiled. 'That's when we saw our marshes returning.'

The Mesopotamian Marshes, sometimes thought to be the biblical Garden of Eden and now a UNESCO World Heritage Site, lie across the floodplain where the lower Tigris and Euphrates come together to form an extensive inland delta. Historically when winter rains and snowmelt at the headwaters caused floods to the south, the marshes would absorb this excess like a sponge, swelling outwards with seasonal growth and then shrinking in the lean summers by draining to the Persian Gulf. The inundations deposited silt from the mountains that fertilised the land, creating a diverse, lush ecosystem in an otherwise arid environment.

As recently as the 1960s, in Abu Hasan's youth, the springtime extent of the marshlands was estimated to be as much as 7,700 square miles, harbouring myriad species of flora and fauna and providing an important stopover for migratory birds on their continental journeys. The wetlands are split between three major areas: the Hammar Marsh south of the Euphrates, the Hawizeh east of the Tigris and the Central Marsh wedged between the two rivers. Abu Hasan's village was in the Hawizeh, thirteen miles from the border where the waters spill over to the Iranian side and are known as the Azim Marshes. 'I was born here, and my father was born here,' Abu Hasan told me with pride. 'In the beginning, we had a reed house, and then a mud one, and then a brick one like this. We've always been marsh Arabs.' He, like many here, claimed lineage to the ancient Sumerians.

But as Abu Hasan's home became more robust, the environment around it was being destroyed. First he recalled British soldiers arriving and drilling for oil, probably in the 1950s. Before the war with Iran erupted in 1980, Saddam Hussein drained the marsh directly behind Al Cheka, and a decade and half later systematically decimated what was left. By 1998, the area was a desert, scorched by fire and encumbered by checkpoints, and the fish and reeds and birds were gone. Over 90 per cent of all the marshes was lost; an area larger than Northern Ireland shrunk to about the size of London.

When they broke down the dikes, there was some hope, Abu Hasan told me, and at least it began to look like a marsh again. But looks can be deceiving. 'It was never really the same. We used to have so many plants to grow and eat. Now we just have two or three shitty plants. The marshes never properly came back.'

It's rare that Abu Hasan uses a boat these days. It's sore on his joints and it made him sad to see what the marsh has become. Instead, he sent his broad-shouldered nephew, Abu Sajad, to take us out to check on the water buffalo. We followed him down a rough embankment to where waist-high reeds stretch out either side of a narrow channel until the shimmer of heat smudged the horizon.

Abu Sajad had an assault rifle slung over one shoulder and prayer beads in his other hand. He stepped onto the bow of a long, narrow, fibreglass boat modelled on a traditional wood and tar *mashouf.* Another relative worked the motor. Salman and I took the middle, our bottoms lower than the surface of the water, knees drawn to our chests. A second boat carried Emily, Hana and Claudio behind.

The ride was fitful and the water shallow enough that buffalo walked on the bottom rather than swam. Where they congregated, they had a bad habit of creating earthen mounds that the boat scraped over, and we stopped regularly to lift the engine. There was no shade, and no relief from the clawing heat. Mayflies and mosquitos swarmed us, getting smeared onto moist foreheads. We passed a herd of pink noses and darting eyes, and I was envious of the buffalo hidden and cool beneath the surface of the water. The animals were Abu Sajad's, and all were accounted for.

Other waterways became visible, coiling off into the reeds, but most were so small and shallow that even our boat was too unwieldy. Some days, Abu Sajad went bird hunting through the fissures, but for that he punted in a one-man canoe. We were boxed in by reeds and sludge. 'This is it,' Abu Sajad said. There was open water somewhere beyond, but it was no longer contiguous or accessible from this marsh. We turned back, trapped in a single, claustrophobic channel.

Today, the marshes cover less than 50 per cent of the area they did half a century ago.* There were a few promising years after 2003 when coverage increased, but progress stalled after 2006, declining for a couple of years and eventually plateauing out before the heaviest rains in a generation in 2019 offered a temporary reprieve. The population never recovered either. From a high of half a million in the 1950s, the number of marsh Arabs living in and around the wetlands fell dramatically, with the majority forced out to the suburbs of Iraqi cities or to flee to Iran. Some never returned. Most of the quarter of a million or so who did now live in villages on the fringes of the wetlands or in nearby towns and cities, rather than in the marshes themselves as their parents and grandparents once did.

We got into a pickup truck to follow a single-track road around the dried-out area. Flanking the narrow strip of cement were occasional rusting sections of industrial water pumps from the 1990s, now stranded and obsolete, abandoned to the dusty expanse that they helped create. Ten miles on from Al Cheka, where the marshes began again, we stopped with a young boatman who has been recommended to Salman as the best person to take us deeper into the Hawizeh. At twenty-seven, Abbas had only ever known the hard times here.

He lived in a small unfinished brick house with his wife Hawra, their three children and seven other relatives. We

---

* Although there is seasonal variation, this number will almost certainly have dropped in the time between me writing this and you reading it.

pitched our tents outside, but the temperature never dropped below 30°C. It was a restless night, and the sky burned orange from flare stacks at a nearby refinery, encroaching on the marshes. At 4.30 a.m., in semi-darkness, we peeled ourselves from the nylon and stumbled into the boat Abbas had prepared. He gunned the engine along a short canal and out to open water. A perfect amber orb rose over Iran and Abbas carved towards it. Soon we were on a glassy lake lit by nacreous dawn light. It was oceanic in scale compared to anything we've seen so far. When Abbas cut the motor, for the first time in many months of travel I couldn't hear a thing. Wilfred Thesiger wrote of 'the stillness of a world that never knew an engine'.* Briefly, our imaginations drifted there, too. Beneath us, algae swayed gently like a forest canopy in a breeze.

'This is the purest water anywhere in the marshes,' Abbas said. Carp darted around and kingfishers and gulls flew overhead. There are at least fifteen species of fish in the Hawizeh, and over 150 species of birds found throughout the marshes, including four endemic and eight that are globally threatened. Some, like the Basra reed warbler, survived the draining of its habitat against the odds, breeding in pockets of remaining reed bed before returning after the re-flooding.

We docked by an island called a *chubasha*, formed naturally by layers of decomposed vegetation. Heavy reeds grew aggressively, some twice our height. The thicket was only broken by occasional narrow passages that materialised like leads in sea ice, inviting but precarious. The landscape was one of concealment, better at masking than revealing what

---

* Wilfred Thesiger visited the marshes each year, usually from late winter until summer, between 1951 and 1958 with only one exception, in 1957, sometimes staying as long as seven months. His book *The Marsh Arabs* (first published in 1964) brought the wetlands to a new and large audience, and reading it was the first time I encountered them.

lay within. Two elderly fishermen, Abu Fayal and Abu Hussein, were unloading the morning's catch in a small clearing on the island. They had come from Amara, an hour's drive away, and tirelessly tossed fish from boat to box, then cleaned the nets of scales and algae. Catfish, not considered halal by most Muslims, were thrown back. Binni, short but deep-bodied and endemic to the region, were afforded good care as the most valuable.

Abu Hussein's boat had no engine and had to be rowed or punted. I briefly wondered if one way to make the marshes feel large again was to slow the pace at which you move, but that was romantic nonsense. It was 40°C in the shade by 8 a.m. 'It used to be perfect weather at this time,' Abu Fayal said. 'But now it's hell on earth from May to September.'

Other fishermen in the area used the same high-voltage transmitters to electrocute fish that we'd seen elsewhere, killing eggs, bottom feeders and anything else in the vicinity. It was illegal here, too, but widespread and rarely punished. 'There's hardly any fish already, and these bastards kill the rest,' Abu Fayal said bitterly of the electro-fishermen. 'In Saddam's time, he would have hanged them.'

Saddam was no friend of the marsh Arabs. For much of his time in power he persecuted them, and a major reason for draining the marshes was to root out Shia revolutionaries and dismantle any ability to rise up. So many communities in Iraq were persecuted by Saddam Hussein, and yet across the country I often heard backhanded compliments of Saddam Hussein's regime. This happened in Kurdistan too.

Sometimes it was outright nostalgia. It's true that initially his reign saw dramatic changes in Iraq. Oil production boomed, and during the 1970s and 1980s there was a concerted effort to eradicate illiteracy by offering free education, to subsidise food and housing and to haul the standard of living up nationwide. There are young people today who have never known stability and assume that Saddam must have been doing something right then. Older people, like Abu Fayal, sometimes lamented the breakdown of law and order, and the lack of a clear, robust system. Saddam, the brutal dictator who massacred his own population and started wars that crippled the country, still attracted a resentful respect from some of those who had suffered most, and that grudging admiration only grew the farther Iraq fell from grace.

Abbas estimated that there are only maybe a hundred people still making a living from fishing in the Hawizeh, and no permanent inhabitants. It had been the most resilient area during Saddam's time because it is fed by the Karkheh River in Iran as well as by the Tigris. But in 2001 Iran started constructing a forty-mile-long embankment to stop water from flowing to Iraq, and now much less comes down the river. Then there is the Ilısu Dam, the full impact of which is still being measured. For Abbas, the future was not bright, but he was going to remain. 'For as long as I can,' he said, 'because what else would I do?'

Chapter Thirty-One

Uncertainty

Days 62–65
Chubayish | Central Marshes | Hammar Marshes
River miles: 1,050

Unusual things are said to happen in the narrow channels of the marshes. We heard stories of a race of giants, whose existence was suspected because of curious tales of large foot-prints and oversized bones. 'There are *djinn* here,' we were told: ghosts. The people who lived in the marshes were much more open to the idea of *djinn* and giants, said Salman. There was lots to learn from the marsh Arabs' philosophy, as well as their knowledge of the area. Did I know, he asked, that not only did each type of reed have a name, but each stage of its growth had a different term, too? Salman knew someone who was collecting all of this, including words from the local dialect, so they would not be lost.*

We stopped at a lean-to where fishermen were cleaning their nets. I lay down, wiping sweat from my eyes, and stared at the sky. During that time, while I was occupied with

---

* The man doing this is called Ahmed Saleh Neema, who shares all of his work on YouTube and social media. Separately, the Nahrein Network in London have also supported a project to compile a dictionary of the marsh Arab dialect.

self-pitying thoughts about the heat, one fisherman produced flour, water and salt from his bag and began kneading. Then he gathered into a pile the long green leaves of the young reeds and placed the dough on top. He next collected a bundle of *qasab*, the thickest and most common reeds, and set the end of the bundle alight so it burned like an oversized collection of incense sticks. He slowly rotated the smouldering bundle as sparks crept along the reeds. Within fifteen minutes he had baked bread for breakfast and was soon frying eggs in a small pan. We were welcomed to join.

There were still traditional techniques being used in the marshes, like this way of making bread. Earlier we'd seen a five-pronged spear with a long handle called a *faleh*. A man stood on the bow with the spear held firmly in both hands, watching for movement in the algae as the boat crept along, then suddenly he thrust it down, making a split-second calculation for the refraction of the water. It was majestic.

Tourism was one way to help protect all this, said Salman. He knew it was a model that had been a saviour for traditions elsewhere, and the industry could subsidise lifestyles that were otherwise struggling in a modern economy. There were a few individuals who had already tried to start this, in a limited way. It was ambitious, of course, and would require more stability both politically and environmentally. But the idea of a sustainable tourism project in the marshes was an exciting one.

Although the three parts of the Iraqi marshes are no longer contiguous, we made our journey through them as one, connecting by road when the wetlands ran dry. The biggest town was Chubayish, gateway to the Central and Hammar Marshes. It had grown to become almost unrecognisable in recent years, said Salman. Once it had been like a second home to him, where he first came to learn and care about

Iraq's environment, and he still counted many friends among the population. We stayed in the home of Abu and Um Haider, who I also knew from previous visits. Abu Haider was a small man, with a face carved by a lifetime of fresh air, and he brought us into his guest room while his sons made tea. A large air-conditioning unit in the corner hummed, and we gently vied for the closest spot.

It was another long night. The electricity cut out around 10 p.m. and did not return. There was nothing cold within reach to cool us down, so the only solution was to breathe deeply and purposefully, and to make sure not to move unless absolutely necessary. I filled the rest of the time by questioning why on earth I was here through choice.

Abu Haider guessed they got about eight hours of power a day if they were lucky, and the rest of the time they had to make do. Before long it would be 50°C here in the middle of the day. The hotter it became, the more the infrastructure struggled in Iraq. The same was true for people, of course. At that temperature, it hurts to breathe through your mouth so you used your nose. You can feel your eyeballs drying up and when you blink it scratches. If there is hot wind, like a hair-dryer, it makes it worse. If there is humidity, you're done for. I found myself sipping water constantly otherwise my throat would swell up. It's too hot for schools, and anyone who can stays home, trying to touch the fewest possible surfaces and praying for the electricity to kick in.

We walked under moonlight to the jetty at the end of the street, passing a large traditional building, a *mudhif*. It stood long and graceful, like a cathedral of reeds. These were the zenith of marsh architecture. Mostly they were owned by sheikhs and used to host community gatherings. Their importance had been diminished in recent decades as concrete villas sprang up around the marshes, but they were still potent symbols, representing the unique way of life of the marsh Arabs and the influence of those who could afford to build one. Inside the marshes, when we completely left solid earth behind, we'd see more humble versions called *serifas* that were still used as homes.

Abu Haider's boat was, like the ones we'd used in Hawizeh, based on the traditional *mashouf* design, but machine-built now from fibreglass and wood panels and powered by an outboard engine. A long pole was still stowed inside for punting through reeds. We clambered inside and Abu Haider produced floral blankets to sit on.

In the half-light of dawn, grey haze fused earth and sky and the surface of the water lay still and even, cracked only by our movement. Abu Haider began to sing. It was a tune to a scale I did not know. Perhaps neither did he. We were flanked by thick green reeds, glittering in dawn glow. Occasionally a water buffalo would ease its sleek body below, until only pale horns and pink nose were visible. Another boat passed with a bellyful of bundled wood, sailed by a woman all in black. Water rushed by the bow, an orchestra of wind and reed, in time with the rhythm of a bird's wings as it landed atop the glass. We'd drifted to another world, where little had changed for five millennia.

Soon the reed beds were aflame with amber light. I looked over the side of the boat, and like a mirage a Euphrates soft-shell turtle floated by, its leathery olive back wrinkled by the water. I had been hearing about them since Turkey. I saw this one only fleetingly, but it was as beautiful and important to me as Taq Kasra or the Malwiya minaret. It filled me with hope, that not all was yet lost; still there was a chance to drift peacefully past a turtle in the marshes.

We came, in the middle of the marshes, to a long, straight strip of concrete. It sliced across our path, disappearing into eternity both left and right. We walked upon it, and I felt its solid weight beneath my feet. 'This was Saddam's road,' said Abu Haider. 'After he drained the water and life from this place, he built roads so his armies could move around more easily.' He lit a cigarette and pawed at the pavement with his sandal. Long, narrow toes stretched out over the sole, and his nails scraped rutted concrete. 'This is all that's left from that time, and now it goes nowhere. It's surrounded by water on all sides.'

There was one place where the concrete had been smashed through, enough for a boat to cross if the engine was lifted. So we crossed the road and made for a collection of reed huts in the distance. This was the village of Yishan Gubba, and we drew the *mashouf* up onto the fractured bank of the *chubasha* where an obstinacy of buffalo lounged in the shallows. I'd been here in 2018 when the drought from the previous year had killed several buffalo and left the herder I met financially crippled. He was on the verge of leaving then, and now I was happy to see as we approached that he was constructing a wood-framed pen for his animals. There were fifteen or twenty young boys with him, and curious young buffalo calves prodding their flat, wet noses into the crowd.

The herder greeted us formally and gave me three short, puckered kisses on my cheek. Things were better, he said. 'Two thousand and eighteen was the worst year I remember. There's a chance this year will be almost as bad. But for right now we have some water, and I've had a couple of good years. The buffalo milk has sold well, so thanks God, we can live.'*

We stayed for three days in Chubayish, and each day made excursions north to the Central Marshes or south across the Euphrates to the Hammar. Abu Haider never needed a map of the marshes, nor were there any that could have guided us

---

* In the summer of 2022, the water around his village almost dried up completely.

as he did. One afternoon he led us through a series of channels in the Hammar until we reached a complex of traditional buildings. The closest was arched, bound by exquisite and intricately woven bundles of reeds and open at one end where the ashes of fire smoulder just inside. The smell of freshly baked bread wafted out to greet us. This serifa was the home of Abu and Umm Jassim, and the other two buildings, also of reed, housed their buffalo.

Emily and Hana joined Umm Jassim inside, and the men sat outside with tea. Aside from a cracked Nokia mobile phone in Abu Jassim's pocket and a generator covered with blankets, there were few other visible adaptations from the traditional lifestyle. Abu Jassim guessed that, unlike in the Hawizeh, there were still a few thousand marsh Arabs living permanently here in the Hammar. 'Our life has always been simple,' he said. 'We just need to be able to look after ourselves and our buffalo.'

Abu Jassim did not share the herder's sentiments from the Central Marshes. The water in his area was almost as low as the last drought in summer of 2018 and what remained was poisoned. Boat engines corroded in the salty water and fish were dying. His buffalo couldn't drink from the marsh any more. Umm Jassim brought us boiled, sweetened buffalo milk and hot bread, and we reclined on the soft reed floor. Her son Jassim played with a calf. 'You've got to earn their trust,' he told us solemnly, then shook his tussle of sunbleached hair and ran inside. I ask Abu Jassim if they'd ever leave. 'I couldn't do that to my wife and daughters,' he told me. 'We can't go. Not again.'

In the last half-decade, the saline level in some areas of the marshes has increased by as much as fiftyfold. This is in part due to increasing saltwater intrusion from the Gulf, which is in turn caused by the reduced flow of the rivers and rising sea levels. The volume of water arriving to Iraq through the Tigris and Euphrates has already decreased over 30 per cent since the 1980s, and the water available in 2025 is expected to be a full 60 per cent less than it was in 2015. Climate change in the region will cause a rise in temperatures at the headwaters which, along with increased run-off and shifting rainfall patterns, will cause a decrease in the available water and further exacerbate the decline.

The United Nations Environment Programme had recently named Iraq the fifth-most vulnerable country in the world to the consequences of climate change. In January 2021, President Barham Salih had ratified Iraq into the Paris Climate Agreement and, on World Environment Day, wrote that confronting climate change must be a national priority. He spoke, too, of returning to Iraq's ancient greenery, heeding the lessons of the past. This was undoubtedly true, but as one leading Iraqi environmentalist told me when I was preparing for the journey, 'climate change provides a nice excuse for covering up bad decisions'. It was used by politicians to cover their bad decisions and inaction. It was a challenge, yes, but it was not the reason why Iraq found itself in crisis.

As we left Abu and Umm Jassim, I asked Salman what he thought. The marshes were the closest place to his heart, he said, 'but what is happening here is also happening everywhere'. The issues in the wetlands were indicative of environmental problems across the whole country. 'Look at what we've seen. Wastewater and pollution everywhere. The government is allowing this to happen to our own country.'

He outlined a list of the dangers. Mismanagement was top. Then there was the increasing temperature, and weather systems becoming more erratic. Decreased precipitation, drought and desertification were adding to water scarcity. Water-related conflict between families and tribes was becoming more common. Factor in an expanding population and the perpetual cycles of war and instability over the recent past, and it is perhaps no surprise that there has not yet been a government willing and able to address the crisis. 'But they should make it a priority,' said Salman, and I thought of Ali Al-Kharki's warning that the water crisis was greater than any threat posed by ISIS.

There was also great potential in the marshes that was not being used, which frustrated Salman greatly. When we returned to Chubayish, he took me to the office of the environmental NGO Nature Iraq to meet its director, Jassim Al-Asadi. Jassim was born in the marshes and had been an authoritative voice on its precarious state for years, as well as a mentor for Salman. 'The Ministry of Water Resources needs to understand the power of the marshes,' Jassim said, speaking from behind a large desk hidden under mounds of papers and books. He paused to switch on the AC, and we collectively sighed with pleasure.

The majority of Iraq's water goes to agriculture, Jassim reminded us, where a lot is lost via archaic water management and irrigation systems. If the marshes were prioritised, he told me, each precious cubic foot would be more productive, and a more productive marsh is important not only for the people who live there, but for everyone. Vast marshes like these are powerful carbon sinks, he said, smiling and tapping the desk with a finger. He, like Salman, also believed in optimism. 'I have a lot of ideas. The government won't follow all of them. But even if they start with 20 per cent, that'll help a lot.'

Chapter Thirty-Two

The Final Confluence

Days 66, 67
Uzair | Qurna
River miles: 1,094

When we returned to the main channel of the Tigris, it felt at first strange to be on a body of water so narrow and constricted. But soon familiarity returned. Fishermen waved from low boats, their nets strung out across the sweep of the river. Kingfishers hovered, the *wireat* were as populous as any flora or fauna, and always the sound of the insects of summer was our accompaniment.

Our boatman to Basra was a boatbuilder, like his father had been. His father had made traditional *mashouf* from wood and tar, but he used fibreglass. Most were thirty-two feet, narrow and long, low in the water. He also did good business making ten-foot canoes for bird hunting in the marshes.

Villages now were hidden behind lush vegetation. Around one, gnarled cottonwood trees left the landscape covered in a dusting of brilliant white cotton. This phenomenon lasted for a month, and in the short time we passed through we, too, were sprinkled with the snow of the south. On the right side, beyond the cotton villages, rose a large cerulean-tiled dome, and under it low buildings of yellow brick. Steps led up from the river. Inside, Hebrew script noted that this was the resting place of the Jewish priest and scribe Ezra, who led Judeans to Jerusalem from their exile in Babylon. Behind a locked door, in a wooden cenotaph, was said to be his tomb. Once this had been the centre of a large Jewish community and a destination for pilgrims. The town that grew around it was called Uzair, the Arabic form of the name Ezra.

Fixed into the walls were Hebrew prayers on stone tablets. But there were also pictures of Saddam Hussein, and green flags hung by the portico entrance. The books in the alcoves were not Torahs but Qurans, because this was also a holy space for the Shi'a. A few other visitors joined. They were Muslims, unaware of the Jewish history. It didn't bother them, one woman said, and it made sense because Jews were People of the Book. But she was just there for the miracles she'd heard about.

Next door was an empty space that had once been a synagogue, and beside it a former religious school. Sleeping in that room now was the caretaker, and when he saw us he rubbed his eyes blearily, adjusted his *dishdasha* and hobbled out to find the sheikh.

'I hear there're foreigners here,' said the sheikh, gliding through the doorway and pressing his prayer cap even more tightly to his head. He sat on the floor and motioned for us to join him. The sheikh's father, a Muslim, had begun working there in the 1940s, just before the Jews were forced from Iraq. After 1950, when the community was almost all in exile, his father was elevated from security guard to manager of the site. He was proud of this. The order to remove the Jews came from Baghdad, he said, and his family had nothing to do with it. His father worked the rest of his life at the tomb, looking after both the Jewish and Islamic heritage. He took over when his father died.

A member of the local *mukhabarat* arrived as we were leaving, alerted to our presence by someone in the compound. He took Salman aside, conspiratorially. The intelligence officer told Salman that the sheikh was known to them. He was an Iranian spy, the man said, with ties to Hezbollah and other militias. That seemed unlikely, and a much too juicy and convenient conspiracy theory. It also seemed to be that, whatever the circumstances that led to the tomb in its current form, it was still preserving some record of the past. If the shrine could truly be shared, then that would surely be something to celebrate.

The Tigris and Euphrates reunite at Qurna, just over a hundred river miles from the Gulf. We dropped into the confluence from the north. The Tigris was small and muddy, and overpowered by the broad, irascible swing of the Euphrates which glowed green under the surface. Beyond, the two became the fluvial superhighway of the Shatt Al Arab.

Where they joined, a smart corniche had been built in recent years. At dusk it was busy with groups of young men and women, out in their best threads to take selfies and smoke shisha. A couple of cafés provided plastic seats where older couples sat, watching the descent of the pale sun. The rivers rippled in a light wind, but otherwise showed no great excitement at the union.

Somewhere under the water here lay a great treasure. In May 1855, four *keleks* were loaded with antiquities from Nineveh, Khorsabad and Nimrud, and prepared for a journey to Basra. The convoy was arranged by the French consul in Mosul, Victor Place, who had excavated many of the artefacts, and a Swiss man with the surname Clement was given the responsibility of ensuring safe passage. He described the contents thus:

It would be impossible for me today to give a detailed account of these masterpieces of Ninevite art; but, among the most striking are two gigantic, human-headed winged bulls, weighing each 29,400 kgs, and two large genies, accompanying the bulls, 4.8 m tall and each weighing 12,700 kgs. All of these, bulls and genies in so-called Mossoul marble, a type of gypsum, were bound for Paris to adorn the entrance of the Louvre just as they adorned the entrance of the palace of the kings of Nineveh.

The museums of Paris and London already had some comparable objects, albeit cut into two or three parts to facilitate their transport on rafts and their transfer onto ships, while these [i.e. the Khorsabad sculptures] were complete, without making any cuts ... enormous expenditure and untold trouble had been required to transport them in such a complete state of preservation.

Alongside these colossal trophies of Assyrian sculpture, over 150 crates of all dimensions contained alabaster or basalt statues; beautiful bas reliefs; art objects in iron, bronze, gold and silver, etc. etc. The cuneiform inscriptions that were found on most of these objects, and their perfect state of preservation, considerably increased the artistic and scientific value of all of these antiquities.

Clement almost ran into trouble a few times on the way down the Tigris, at Kut and Uzair, but was able to pay off local sheikhs. Then, just above the confluence at Qurna, his *keleks* were set upon by tribesmen. He and his crew were forced off the boats and stripped bare on the riverbank, and the raiding party took what they could. Two *keleks* were sunk, and the others drifted crewless to Basra. Clement, once he'd found a pair of trousers and talked his way onto a boat to Basra, lamented the loss. 'Of all of this beautiful collection of antiquities, only one bull, one large genie and twenty crates containing bas-reliefs, escaped the disaster.' He rallied a salvage party, and they pulled another of the winged bulls and a few other small pieces from the water, but the rest was too deeply embedded in silt. There have been several attempts since to retrieve the horde, including a Japanese expedition in the 1970s, but until now the looted antiquities destined for the Louvre remain at the bottom of the river.

Behind the cafés and the teenagers at the confluence was a small shrine enclosing a decrepit, blackened tree that many local people believe to have been visited by the patriarch Abraham (*Ibrahim* in Arabic) four thousand years ago. Others, usually foreign travellers, had claimed it as the biblical tree of the knowledge of good and evil. But if it was sacred, it was also clearly dead. It was not clear what type of tree it had been, though it was not a palm. One of the large branches had been broken, reportedly by British soldiers climbing on it during the First World War, and where the wood had cracked concrete had been pasted in to seal it up. A sign in English and Arabic told us that Ibrahim 'prayed and said here a tree will grow which is similar to our father Adam's Tree, Peace be Upon him in Adam's Paradise'.

We thought the whole thing amusingly dilapidated, and after we were done we went to sit by the water. An old *haji* was stretched out on a rock with a fishing rod. 'We saw a lot of strange things with this tree,' he said, catching our eyes. 'It might not seem much, but let me tell you: it hasn't changed since I was young. And during the war with Iran, its trunk was crying with red tears.' A younger man passing by said he believed, too, and I felt a little bad for making fun of it. The *haji* said he would pray for us, and I thanked him. We could use it for the home strait.

In the morning, we awoke to the sound of seagulls. Seagulls! But we had one final detour to make before we followed them south. On the Euphrates, close to the confluence, was a town called Huwair where one traditional boat workshop was still in operation. We were going to borrow two of their canoes for a day. The workshop was owned by Abu Sajad, lean and bare-legged, and run at the behest of Rashad Salim. When we arrived, Abu Sajad had just upturned a large *tarrada*. These boats were the Rolls-Royces of the south, thirty feet long with a sweeping, tapered stem at the front and pronged with decorative nails along the ribs. No one had built them since the 1980s, but Rashad had commissioned this one, inspired by the *tarrada* given to Thesiger by a marsh sheikh.

For four hundred years, Huwair had been a boatbuilding hub. Until the early 1900s there were still Mandaeans working here. The last of the great boatbuilders in Huwair died only a few days before we arrived. He had made the final functional boat that anyone remembered, which was used to take children to school three and a half decades ago.

Abu Sajad and his team were coating the hull of the *tarrada* with bitumen, and took it in turns to carry pancake-sized piles of molten tar from a bubbling cauldron, then spread it along the smooth wood with rolling pins. It was only him and Abu Kathem, who was overseeing the bitumen application, he said, that knew how to do this any more.

I asked about the process for the *tarrada* and was pleased when Abu Kathem obliged by outlining in detail. There were ten stages, he said, beginning with laying out a blueprint on the ground with string. Then they began with the ribs, followed by the base and the sides. Each stage had a name, unique to the craft, and I thought that here, too, was another dictionary that needed to be compiled. On the *tarrada* the bow and stern were ornate and complex, and required additional steps to make them secure and streamlined. The last step was applying the tar, in two layers, and each of these processes also had a name.

Abu Kathem was generally agreed upon to be the most skilled craftsman here. He was small and broad, with strong, dry hands that looked as if he spent a lot of time in a workshop. Often he began a boatbuilding process with a scale model. These miniatures were as beautiful as any I'd seen. Emily in particular was excited by the artistry, and spent a long time turning the boat over in her hands. Abu Kathem had been making models since the 1980s. In the 1990s Saddam Hussein commissioned one, and only a few months before we arrived one was presented to the Pope during his visit to Ur. Abu Kathem was still waiting for an email to say whether Francis liked it.

'The next generation knows nothing about this,' he told me. 'My son is in university. He said this stuff isn't for him. When my generation is done, it'll fade and then it's over. The youth don't have the will and the passion.' Had it not been for Rashad Salim, it might already be gone. But Salman, as ever, was optimistic, and told them both that he believed in the youth. 'Look how young people are picking up the old skills elsewhere,' he said. 'We can find people to keep this alive.'

We brought two *mashouf* by truck to the Tigris. Boats here were measured in 'arms', and a *mashouf* was seven arms compared to the fourteen of a *tarrada*. Smaller was more manoeuvrable, better for wriggling through reeds, and larger showed status. I got in one boat with Abu Kathem and Abu Sajad, and the others rode with a big jolly man who loved two things: paddling and singing. As soon as we had pulled

the first stroke, he began to serenade Hana and Emily, sur-
prising the seagulls who'd settled alongside.

For a few wonderful hours we moved, me at the bow, paddling on one side, and Abu Kathem behind on the opposite.
I felt my shoulders creaking, my forearms tightening: I had
desperately missed this physicality. It was 49°C, but there was
a sweet spot where, with just the right amount of effort, we
could create a cooling breeze but not overexert ourselves. The
river passed slowly, and in the *mashouf* the bottom half of my
body was below the waterline so I felt even more connected
to it. This was how I had envisaged the trip and although it
would never have been possible, I enjoyed imagining what it
might have felt like. If there was no threat from ISIS or the
Hashd, and no grave mines or dams, how would that have
been? But, then, that would not have been the Tigris. So
instead I thought of the movement, the here and now, and
the sound of water on tar, the slice of the paddle, the singer's
vibrato. And just like that, I was ready for the Gulf.

Emily paddles a homemade raft through a rock arch at the source of the Tigris in Birkleyn. There is a theory that the complex of openings at the source are associated with a cuneiform phrase which can be translated as 'divine roads of the earth'. These metaphysical paths are gateways to the underworld. Claudio von Planta

The team for the Turkey section of the journey, pictured above the Dicle Dam where the two arms of the Upper Tigris join. From the west, the river flows in from Lake Hazar, and to the east it comes from the Birkleyn caves, where we began our journey. Left to right: Angel, Bişar, Claudio, Emily, Leon. Claudio von Planta

The Tigris River in the mountains of south-east Turkey, close to a village called Sulak. Below, nomadic shepherds move their herds of goats along the flanks of the canyon. Just to the south of where the picture was taken, the river spills out into the city of Cizre, close to the borders with Syria and Iraq. Leon McCarron

A homemade *kelek*, constructed by a man called Mesut at the confluence of the Tigris and Batman rivers. Four tractor-tyre inner tubes gave it buoyancy, and Mesut had carved a small seat for the rower. Mostly it was used by villagers to cross to the north side of the river where they could then go by car to Batman for work or shopping.
Leon McCarron

Emily taking a photograph in the church inside the monastery of Mor Augin. It may have been built as early as 361 AD. Much of the plaster is still original, applied by some of the earliest adopters of Christianity. Leon McCarron

Claudio filming from atop the Zangid Bridge in north-east Syria. Once the Tigris rushed underneath, but now it has been rerouted and the bridge stands alone, close to the international border. As the team climbed the bridge, we were warned that until recently anyone standing on top would be shot at by Turkish border guards. Leon McCarron

Camping on a small island in Mosul Lake, in the Kurdistan Region of Iraq. Many nights on the journey were spent sleeping in these tents, though rarely was the view as pleasant as this. Leon McCarron

A drone shot of the two-mile-long rock wall of the Mosul Dam, holding back four hundred billion cubic feet of water. It has been called the most dangerous dam in the world because of the instability of its foundations, and the threat that poses to Iraq and its population if it were to breach. Claudio von Planta

Emily: It was our first day travelling downstream with Omar under the destroyed bridges in Mosul. He always held a stoic gaze. We would be shot at within an hour of taking this photo, and Omar would protect us. Emily Garthwaite

Emily: Khitam found me in the fish market, asking for a photo or two. It was early in the day, and the market was filled with men and moustaches. She danced in plumes of cigarette smoke and heckled the men around her. Then she took me to her old home, destroyed during the war. She didn't dance there. Emily Garthwaite

Emily: The bodies of two drowned Moslawi children had just been carried into a vehicle, with their uncle following behind. These brave divers, led by the River Sheikh, stood in silence, and the sun set moments later. We left shortly afterwards to photograph families playing in Mosul Forest. Emily Garthwaite

Lunch break at Hamam Al-Alil, where we spent a night in a semi-abandoned water treatment plant. Because of the challenging security situation, we were accompanied by soldiers from the Iraqi army, who travelled along this section of river with us for protection. Leon McCarron

Claudio and an Iraqi soldier in Hamam Al-Alil. The soldier was a member of the Crocodile Club in Baghdad, an exclusive group for men with exciting facial hair. They meet regularly to do community service and compare moustaches. Claudio von Planta

Emily: The children became increasingly excited by a foreigner photographing them diving into the sulphurous waters in Hamam Al-Alil. At one point my camera was hit by a particularly big splash, and the children proceeded to offer their help, their chins and hands dripping more water onto my camera. Emily Garthwaite

I spent the morning with Sheikh Aziz's wives and female family members. Upon returning, he invited me alone into his guest room. He removed his gold silk robe and placed it over my shoulders, and directed me to sit under a rug emblazoned with his face so he could take a picture. Then we took one of him doing the same. Emily Garthwaite

Emily: Salem spoke with us in the most beautiful classical Arabic, and loved the site at Assur so deeply that it was infectious. We spent hours wandering the site with him, but he was still terribly disappointed when we said we had to leave. He made us promise to return and so I did, a year later, for Assyrian new year with Assyrian friends from Erbil. Emily Garthwaite

Chapter Thirty-Three

Water Martyrs

Days 68, 69
Basra
River miles: 1,140

On the approach to Basra, a dust storm kicked up and covered the sun. We wrapped ourselves in scarves and hunkered down low in the boat. The Shatt Al Arab was broad and quiet, and our boatman stayed close to the west bank where buffalo wallowed. Salman had received a call to say there was tribal fighting on the road near the river, which began just a few moments after we'd left. Apparently, it was over a water supply. We should be careful, he was told. It all felt rather ominous.

At one time, bull sharks came up the river from the Gulf. During the Second World War, there were accounts of British soldiers stationed at Ahvaz and Basra who noted a handful of shark injuries and resultant amputations. Even earlier than that, in 1839, a bull shark was caught in Baghdad, hundreds of miles from the Gulf, and its head is preserved in a jar at the Natural History Museum in London. Our boatman told us that mostly local stories of sharks have disappeared in recent decades, but there were still some Basrawi fishermen convinced that attacks still happen. He didn't give the idea much credence, but it encouraged me in my conviction to avoid swimming.

There was little to see, especially from such a low vantage point. One risen mound looked like an ancient tell, but actually it had been built to give Saddam's snipers some elevation. That seemed to sum up this area. In the distance, through the brown haze of the dust, now-familiar flare stacks burned a hole in the horizon. Seventy per cent of Iraq's crude oil comes from Basra, and we were never far from the impact of the industry. Every family here had someone who became sick from the chemicals that flaring released into the air, said Salman.

We had been sleeping badly recently, because in tents it was relentlessly hot, and inside homes the air conditioning was always patchy. Iraq's electricity provision is, at the best of times, notoriously poor, and a constant reminder to its people that the government cannot even supply them with basic necessities. Yet part of the solution, ironically, could be found in the flare stacks.

Only Russia flares more natural gas than Iraq. But Russia, like all the other countries with large volumes of gas released by oil production, is a net exporter. Iraq does the opposite. It both flares off its own supply and simultaneously imports gas from Iran that is burned for power. This is costly not only financially, to the tune of seven billion dollars a year, but also environmentally, and creates a dependency on its neighbour for an essential service. That, of course, is the point, and is one clear demonstration of Iranian influence over Iraq's political system.

If Iraq could capture the gas flared in Basra Governorate alone it could dispense with imports. Previous governments talked of solving the problem, and future ones will have to.

It's not difficult, and the investment required would be paid off relatively quickly. In the meantime, Iraqis survive with terrible service from the grid, especially in summer, and are vulnerable if Iran reduces the amount it exports. If individuals can, they pay extortionate sums for private generator usage to fill the shortfall. Rather than going to the ministry where it could be reinvested in infrastructure, their money instead ends up in the pockets of the generator providers, who are often connected to militias and other shady groups. It was, like so much in Iraq, frustrating to see a problem that impacted on so many and for which there was a solution, and yet in which one administration after another failed its people and left them sweltering in the dark.

We passed a power plant, half crippled after 2003. Pylons destroyed in war lay crumpled where they'd fallen, occasionally with newer replacements built around the debris. But most noticeable of all was an absence. Once, the Shatt Al Arab had been flanked by enormous groves of date palms that ran all the way to the Gulf. Iraqis had called Basra 'the Black City', and the country produced 75 per cent of the world's dates. Then, during the Iran–Iraq War, the trees were uprooted or cut down to make room for a front line. Military roads were poured over the graveyards of the palms, and the boulevard of the Shatt Al Arab was stripped bare.

A haze of dust hung in the air. The gusts came from the east, which made it 'humid in the marshes, hell on the river', said the boatman. It was only the northerly wind that was welcome. The river turned to lead, small ripples turning white and angry and coming at us broadside. We arrived into the city through a graveyard of hulking, rusting ships. It became like a gauntlet, and we had to slow and navigate around skeletal hulls chained to the docks as well as those which had slipped under the water. Now Basra's deep-water port was a place called Um Qasr on a canal that ran to the estuary, but, once, these ocean-going ships we passed had come up the Shatt Al Arab.

I looked up at the names of the ships: *Al Zubair, Duhok, Teiba, Al Mansur*. Some tilted awkwardly on their sides, soon to slide into the Shatt Al Arab like the others. Beneath them, small fishing boats plied their trade, floating under hulls like gulls alongside a trawler. There were large floating docks, too, registered in Zanzibar, and on the bank dozens of decommissioned cranes that had once pulled shipping containers to dry land. Now their rusted mechanical fingers were frozen in a permanent arthritic grip on the poisoned air.

Basra is the end of the line. Everything that found its way to the Tigris upstream – the industrial, agricultural, human and animal waste – eventually ended up in Basra. Water quality is appalling. Treatment plants are not sufficient to make water safe for drinking, and as elsewhere there is no enforcement of the regulations that should stop river contamination. Around three hundred thousand of Basra's residents are not connected to the water network, and many illegally tap into pipes. Many of these people are the pastoralists who have given up on the land, and moved to makeshift settlements on Basra's periphery. This is existential migration, and with each passing year more leave their rural homes. It is also a vision for the future of Iraq, too, on a much larger scale, if things do not change.

In the summer of 2018, at least 118,000 Basrawis were taken to hospital with gastrointestinal illness caused by poisoned water. Protestors took to the street to demand better quality of services and were met by hostile government forces and militia groups. Fifteen protestors were killed and many more injured. The dead were called the Water Martyrs, and the loss of life that summer can be seen as a direct result of decades of failure of the government to provide drinking water to the four million people of Basra Governorate.

We were greeted in Basra by Ameer, who volunteered for *Humat Dijlah* and worked at the port at Um Qasr. It was Ameer who, six months earlier, had met Emily and me in Basra and illustrated so clearly how horrific the situation was. And it was his words about his son, Mo, that impelled us in part to make this journey. Surely nothing could be so bad it would make one reassess their decision to have a child?

At Ameer's home, Mo ran around in his jungle animal pyjamas. He was five, and a bundle of energy. He had hyperactivity disorder, said Ameer, and didn't always know when to stop. Ameer and his wife took turns with Mo as he brought us toy cars, then a tricycle, and then began forward rolls on the sofa. Emily was great with children, and Mo took to her quickly. She chased him around the room and Hana indulged a game of hide and seek despite limited options, and soon he was giggling and rolling on the floor.

'Basra is an amazing city,' said Ameer, over a spread of *masgouf*. He'd brought the fish specially from a place close to the Gulf for our visit. Emily didn't have the heart to tell Ameer she couldn't stomach it, so she picked at the salads with Hana. Emily had lost a lot of weight and had three bad bouts of food poisoning since we left from the source. We were so close to the end and counting down the hours, yet also trying not to wish away the final miles of the Tigris. It was a tough balance to find.

'All the people called this place the Venice of the Middle East. It was full of water, and agriculture and tourism,' Ameer continued. This was the city of *The Arabian Nights*, he said, and the port from where the legendary Sinbad was said to have set sail. It had been an 'intellectual centre of the first order', and 'the focal point of Arab Sea trade which went as far as China', said the *Tourist Guide*. From the Abbasid era it was famed for poets and scholars, but after the Mongol invasions it became a neglected backwater. It was the Ottomans who restored its glory as a trading port, and in the nineteenth and early twentieth century facilitated ever larger shipments north to Baghdad.

Now, said Ameer, everything in Basra was about oil. 'It runs our lives here,' he told me, 'and we should be rich.' Some were – the elite – but most Basrawis saw none of the benefits of the oil money and instead shouldered the suffering. 'Our unemployment here is so high,' he said. It was estimated to be 20–25 per cent, and even higher for youth. The residents dealt with air, land and water pollution, corrupt and inefficient governance, militia control and extortion, and a lack of basic services. 'Everything that's bad in Iraq is at its worst here,' said Ameer.

Mo came to jump between us. Ameer caught him and held him tightly, his large hands wrapped around tiny, bony shoulders. The boy looked up and calmed. Emily asked Ameer again about his family. 'I never regret it, of course,' he began, 'but I worry. That's why I do all this work. Because we should make things better for the next generation. If it doesn't change, no one can live here in ten years' time.'

We took a walk in the old city, whose glory had faded so much as to only be visible if you squinted. In the streets there was still some of the famed *shanasheel* architecture, ornate bay windows with stained glass and fine waxed wood. 'The richest in our city used to live behind these,' said Ameer. Basra had been home to Jews, Armenians, Muslims and Mandaeans, and traders from India and the Horn of Africa. Now it was almost entirely Shia Arabs. There were only a few hundred of the *shanasheel* left in serviceable condition, and the rest sagged over the pavement, listing and collapsing before our eyes.

Through the old city ran a canal, part of the network that had led to the 'Venice of the Middle East' moniker.* Now, the water was stagnant and clogged with garbage. The temperature hit 50°C, and the smell was so overpowering that we couldn't stand it for long. An elderly man passing said to Ameer, 'What a shame for us to have foreigners see this city in this state. I hope they know what it was like before.'

That night, to make some happy memories from Basra, Ameer took us to a funfair. The Ferris wheel overlooked the Tigris and was one of a dozen or more that I'd seen along the river. We tried a series of terrifying rides on creaking, ageing machines, where teenage Basrawis laughed at my contorted face as I was thrown up and down on ponies and in teacups and other garishly painted vessels. We all took a ride on the bumper cars while the young attendant smoked a cigarette and checked his phone. There were no holds barred, and we left with minor whiplash and large smiles. Then, by the river, Ameer had one last surprise for us.

Lined up along the bank, dressed in white *dishdashas* with red and white scarves, was a band ready to play for a small gathered crowd. The musicians were black Iraqis, part of a population of an estimated one and a half to two million, whose ancestors came from the east coast of Africa as early as the eighth century. At that time Iraq was the centre of the East African slave trade. The majority of black Iraqis live in Basra, and Ameer wanted us to hear some of their music.

We sat cross-legged on a woven rug and the band sat too, their backs to the Tigris. Soon, the audience was clapping along. 'This is real Basrawi music,' said Salman. His parents

---

* When I'd once spoken about this with Rashad Salim, he said he preferred to think of Venice as the 'Basra of the West'.

were originally from Qurna, and he thought of himself as a southerner. The style was known as *Khashaba* and had begun in the city in the middle of the last century. Some bands now used ouds and violins but the most authentic version, like this band, relied only on drums and vocals. Eventually the music, like so much from Basra, flowed out through the Shatt Al Arab and spread to neighbouring countries, and was still popular throughout the Gulf countries.

It was dark now, the lights of the city reflecting off the Tigris like ghost ships, and the band drew in closer to the rest of us. The songs were slower now, lower, with lyrics of *djinn* and giants, and above us a few drops of rain from a freak storm began to fall. Emily fell asleep with her head in my lap, and Claudio lay down, too. He had remarkable energy for a man approaching sixty and seemed to cope with the lack of sleep better than any of us. I listened, and my eyes became heavy, and as I looked out over the steely river I hoped that whatever spirits were stirred here would aid us for the final stint.

Chapter Thirty-Four

To the Sea

Days 70, 71
Seeba | Al Faw | Gulf
River miles: 1,210

We went early to the docks to find a boat to take us to the Gulf. Inside the covered market the first sharp, bowed light of day cut through plastic tarpaulins where dried fish hung like ornaments, curled around with wire. Vendors passing through drew them aside like a portière, and worked fast to make displays before the crowds piled in, dealing fish on the tables like cards. Here and there, the rays picked out chosen men.

Ameer asked around, and reported back only one option. It was an old boat from the Ashari area of Basra, designed with wood panelling and a vinyl canopy to look like its ancestors a century ago, but now powered by a growling diesel engine. Normally it took Iraqi tourists on tours for a few thousand dinars. But the owner recognised our situation and used his leverage wisely. It would be half a million dinars to go to the Gulf, he said: $350 dollars. That was extortionate. Usually we paid boatmen one hundred thousand a day, but we all knew we were in a bind. Salman agreed the deal, and we chugged noisily out of harbour.

The boat was loud and smoky, traits it shared with the skipper, Abu Yousef. Shortly after leaving, we sailed over a fishing net which concertinaed underwater and clogged the propellor. After a little swearing, Abu Yousef handed his wallet to Salman and dived off the side, fully clothed, with a knife in his mouth. Soon he burst back up, a mesh of wire in his hand, and we hauled him on board. He took his wallet and went back to the helm, dripping, still swearing.

We rattled under a suspension bridge, arced to allow ships to pass underneath, then past Saddam's old palaces which leaned in towards the river. One was now a cultural museum and another a headquarters for the *Hashd*. A large Iranian flag flew proudly outside the latter, its pole firmly embedded in the Iraqi soil. Beyond the city, the east bank became dry and bare, a far cry from the fields of fruit and vegetables that once grew there. Soon it was a no man's land marked with concrete. There another Iranian flag few, but this time it was on Iranian soil. From this point onwards, the Shatt Al Arab marked the international boundary, and everything on our left side was a different country.

We nestled in close to Iraq, where heavy reed beds disappeared into swamp. There was just one checkpoint, *Abu Filous* – Father of Money – where we repeated the rigmarole of papers and phone calls and conversations in the sun. On the Iranian side, sentry posts were close enough for us to make out the faces of men on duty and, surely, close enough for them to see ours. Then came the port city of Khorramshahr, where port cranes loomed over ocean-going ships that had reached the end of the line. When our engine briefly stalled and we had respite from the grinding soundtrack of combustion and pistons, we could hear the traffic and blaring horns of rush hour in Iran.

The lion's share of the river seemed to fall to the

neighbours, and compact fishing boats with Iranian flags and names like *Al Mahdi* darted around the breadth of the river in a way that the Iraqis' did not. A couple of these pulled alongside us, curious fishermen with weather-beaten faces looking in under our vinyl cover. *Salam*, they said. *Chetori?* It was amazing to me that we, in an Iraqi boat, could have a conversation with Iranian nationals, who would then return to their side of the Shatt Al Arab to go home. One man even handed us an orange.

Ever since we'd left Basra I'd been concerned that we weren't travelling fast enough. We'd been averaging ten knots an hour, and when I made a quick calculation it seemed clear that we'd fall at least ten miles short of the end before nightfall. When I mentioned it to Abu Yousef, he was wholly unconcerned. He'd spent thirty-five years at sea, he said, and ten piloting this boat around Basra. We'd be fine, he repeated, right until the point at which we reached an Iraqi village called Seeba. There Abu Yousef looked at his watch, swore and told Salman we'd have to do the rest in the dark.

At a rickety police checkpoint, Salman and I decided we should spend the night there and go again in the morning. Attempting to reach the mouth of the river in darkness seemed risky and unfulfilling. Ameer and I helped the others off the Ashari boat with the bags, and Salman stayed with Abu Yousef to negotiate. He offered half the money for half the distance. That was received poorly, so he upped it to a quarter of a million dinars, considering the time needed to return.

It took a while for their raised voices to reach us. Abu Yousef was shouting and standing over Salman who, still sat on the wooden bench, looked surprised and a little helpless. 'You're a thief,' shouted Abu Yousef, and for the first time I appreciated what a big man he was. 'If you won't pay, I'll take you with me.' He hit the throttle into reverse, and the boat edged astern with Salman hostage. I stood, dumb and still. There was no such inaction from Ameer. He vaulted the stone wall, hopped onto the concrete flood defences and launched himself across open water to land rather perfectly on the bow just before the boat reversed out of reach. It was the coolest thing I had seen anyone do in three months.

Ameer was a big guy, too, and he confronted Abu Yousef. They squared up to each other, though Ameer kept his arms by his sides. It did not seem like a good idea for anyone to get into a fistfight on a boat idling midway between Iran and Iraq. The shouting continued, and Abu Yousef pulled the boat back further still until it was firmly in the middle of the Shatt Al Arab and the voices lost.

For twenty minutes we watched until eventually the boat returned, and Salman and Ameer jumped off. We hauled them onto the bank and Abu Yousef turned the Ashari upstream, still shouting over his shoulder. Salman looked shaken. Abu Yousef had accused him of being an American spy, he said. 'He told me, "I saw your cameras, you're American, and you're spying on Iran for them." So he said he'd take me to Iran and let them deal with me unless I paid in full.'

It was hard to know how serious the threat was, but neither Salman nor Ameer could be sure he wouldn't go through with it. The likelihood was they'd all have been shot before the boat reached the Iranian dock. In the end, Abu Yousef accepted three hundred thousand, but said he'd take us to court and send a letter to the Iranians. Finally, Salman smiled. 'I'd like to see them when they receive the letter,' he said, giggling now. 'No holidays to Iran for me, I don't think.' The worry was broken, and we were safe, but still twenty-five miles from the sea.

We slept in the house of a fisherman and in the pre-dawn met a new boatman called Abu Ali. He owned a speedboat and wouldn't move for less than half a million dinars. But unlike Abu Yousef, he was friendly and controlled his boat as if it were an extension of his body. This man's a professional, I told myself, and his boat was sleek and new and well-looked-after, which was what we needed. Emily and I discussed the cost and agreed. The last weeks of the expedition had taken us over budget, and now we had each racked up a sizeable personal debt. But there was no choice, and a swift trip to the end would be a relief.

We had travelled barely five miles when the engine blew up. Abu Ali had us bouncing along at forty knots an hour when it happened, and the jolt sent us flying across the small deck until we were all piled up in a corner. After a cursory inspection, the ignition was turned again and the engine coughed halt-ingly into life, but it could only move at a maximum speed of ten knots. By this point, that was a very familiar speed. When I asked Salman what was wrong he translated simply: 'it's fucked.' We crawled along, past the Iranian city of Abadan, aiming for the last town on the Iraqi side. It was 50°C. We'd had two hours of broken sleep, had no breakfast and ten weeks of travel behind us. Emily had stomach cramps, and we were all very much in danger of losing our sense of humour.

The jetties at Al Faw were knitted with the tall, flat sterns and angular prows of sea-going wooden dhows. Abu Ali took a spanner to his engine and we took shelter in the fish market. Emily found a bathroom, and when she came out an old man hurried over with a jug of water to pour over her hands. Even here, at the end of the river, at the end of the trip, at the end of our wits, were the acts of hospitality which had first made us fall in love with Iraq.

I spoke English with a couple of young boys slicing a tuna. One wore a T-shirt that said, 'I support the right to arm bears'. *Haji* Claudio filmed the fish, then went to sleep. Everyone else sat in the shade praying for good news. Predictably, before long, we were visited by two bored, emotionless policemen, who told us that even if our boat was fixed, we couldn't go any further. 'It's a secure area,' they said. 'You need permissions.'

By now we had been in enough of these situations to know the score. So many men in these positions of minor power considered themselves the ultimate arbitrators of the law, and the taller of these two was no different. He refused to listen and walked away to cool himself off by a freezer. The shorter, younger man, clean-shaven, waited quietly. Emily dragged herself from the shade of a fish stand and took Hana and the policeman to one side. She gave an impassioned speech about our journey, and the months of travel, and the hospitality of Iraqis. 'We've come from the Turkish mountains to be here,' she said. 'Do you want to be the one to stop us now?'

Perhaps it was her way with words, or the look in her eyes after days of dehydration and gastroenteritis, but he gave in. 'You can go,' he told her. Then to Ameer, 'but only to the last Iraqi checkpoint, otherwise someone will shoot you'. Ameer grinned. We should have been fed up with being told that we'd be shot, but this really felt like the last time. The final checkpoint was at the mouth of the river and suited us just fine. I was delighted to have been proved wrong, that petty bureaucrats were not all the same.

Abu Ali had jerry-rigged the engine and, after Emily made one last visit to the outhouse, we powered into the Shatt Al Arab. Opposite Al-Faw, in Iran, was a large concrete monument built to commemorate the martyrs of the war between the neighbours. This area had been devastated, and the cities of Khorramshahr and Abadan completely depopulated. And there, where the coast became a delta, and where countless young lives were lost in a brutal and needless war, Iranians gathered to pay their respects.

It was also there, seven years earlier, that I had arrived at the end of a journey along the Karun River in Iran. I had travelled with a filmmaker friend, Tom Allen, who was married to an Iranian and spoke good Farsi. We'd walked to 11,500 feet in the Zagros Mountains in February to find the source, then hiked downstream through blizzards and five-foot snowdrifts along the trickling waterway. After a couple of weeks on foot, we inflated packrafts that we'd been carrying in forty-kilo backpacks, and paddled downstream into spring and foothills. Eventually the dams on the central section of the river took us off the river and back to the road, so we borrowed a couple of bicycles from a stranger, promising to return them when we finished, and cycled to the southern city of Ahvaz. Then, for reasons I can't quite recall, we ran from there to the river, through the desert, and finished at the martyrs' monument under the gaze of an enormous poster of the two ayatollahs.

I'd made that journey when I was younger and had a higher tolerance for risk. We had no permissions and were often stopped by the police. Sometimes they found us on the river and took us by car to the station to question us, but usually it finished with them buying kebabs and offering us a place to sleep. I'd flipped my packraft in whitewater and got trapped underwater, and we'd camped alongside the river in a protected nature area that turned out to be a bear reserve. All of it seemed desperately careless in hindsight.

It had differed greatly from this journey on the Tigris. It was probably more fun, I thought to myself now, or at least more adrenalin-filled. But Tom and I didn't have the resources to learn much about the Karun, and we were too nervous to ask people we met about the state of the river. So it became an adventure, and the film we made showed in broad brushstrokes an Iran that our audiences might not be familiar with.

What we'd done here was certainly an adventure, too, though in a different way. I would remember it for the rest of my life. I'd shared it with Emily, and the team, and above all I was proud of making the journey alongside Salman and Hana. If they could carry some of the experience of this into their activism, it would all be worthwhile.

The yawning Gulf broadened until the horizon was a pale blue that met the sky. Abu Ali checked his GPS. 'We're here,' he said. 'The end of the river.' I looked around at the fishing boats coming and going, and the swaying reeds. Ameer was the first to jump in, fully clothed, and the rest of us followed. At first, my feet dragged on the bottom, but as I swam out, my legs flailed in open water. I ducked my head and came up tasting salt.

Emily swam over and we hugged, and soon everybody else was there, too, and we held each other in the water. From the boat, Abu Ali passed a bottle of water that I had stored at the bottom of my bag. I'd filled it from the Tigris tunnel and now, in the Persian Gulf, I removed the cap. We each took a swig. 'It's holy water,' said Salman, and I poured the rest ceremonially back into the river. The last drops disappeared and drifted downstream past the final checkpoint alongside the grains of mountains carried from the north. Here was the final act of the Tigris; the dénouement of the rivers that shaped Iraq. 'Let's hope,' said Salman, 'that these drops are not the last ever freshwater to travel down the Tigris.'

There are several Sumerian and Babylonian myths that deal with the creation of the gods and of the human race, and in all of them water plays a central role. In the Sumerian myth *Lugal-e*, the creation of the Tigris is described. A warrior-king, and god of rain and the spring flood, Ninurta, is called to do battle with a new challenger in the mountains. After fierce fighting, Ninurta is victorious, and decides to reorganise the landscape in celebration. He collects the annual floods, previously spread throughout the mountains, and by reshaping the foothills into channels he directs them to the Tigris to inundate and irrigate Sumer.

The gods of ancient Sumer were often interfering with the Tigris in different ways to ensure it provided fresh and plentiful water to the early city-states. Now, at the end of our journey, after following a river mined with shallows and threatened at every bend, I thought of the words of *Lugal-e*, which described what it was like before Ninurta's intervention: 'The Tigris did not bring up its flood in its fullness. Its mouth did not finish in the sea, it did not carry fresh water.'

It seemed very familiar. On our journey, when we asked about the future of the Tigris and what the solution was, or who could help, one of the most common refrains we heard was 'Bas Allah'. Only God. The river today is facing an existential threat, and those who rely on it are looking to the heavens for help, just as their predecessors did for millennia. Who am I to say that won't help? But it had also become clear to me that the villain was not a monstrous warrior or demon in the north. It was the avarice and carelessness of mankind, and if that didn't change, from source to the sea, then it was certain this river would dry up, until all of Mesopotamia and Iraq became a lifeless desert, and there was no god that could help it then. And, if it could happen here, at the very birthplace of civilisation where water management was pioneered, then it could happen anywhere.

We climbed back onto the speedboat, our clothes already almost dry in the sun, and Abu Ali spun us around, signalling to the watching sentry for the last time that there was no need to shoot us. Behind us the waters of the delta continued to agitate, and the Tigris crawled onwards until it became lost in the Gulf.

Author's Note
and Further Reading

I have tried to be as honest as I can in this book, but it is perhaps worth adding a couple of notes on my process here, for clarity. All of the characters are real, and our interactions described to the best of my ability. None are composites. There are a couple of people who have had their names changed, and similarly one or two whose locations have been slightly altered, to protect their identity. Otherwise, everything you read is my attempt at representing people and places accurately.

As we travelled I carried an A5 Moleskine notebook and transcribed conversations longhand. I'd usually record them on my phone too, if that was possible, and take reference pictures. For some interviews, Claudio would be filming. When I came to write up dialogue, I relied primarily on my handwritten notes, then cross-referenced them with the recordings. Most of the conversations were in a language other than English, and the translations – from Angel, Sam, Salman, Hana, Ali or anyone else – inevitably have a little of the interpreter's own personality. For anything that seemed very sensitive or confusing, I usually had someone else check the original recording if there was one.

I recorded all our movements digitally. I separated the GPX files into road and river travel, and this is how I came to the numbers of river miles that accompany each chapter. By the end, my data suggested the river had run 1,210 miles from Birkleyn to the Gulf. That's a little longer than most other estimates (an average seems to be around 1,180 miles), but I hope you'll take those figures in the spirit that they're intended, as a rough guide. I reckon we were actually on the water for around 820 miles. That's 68 per cent of the journey, which, given the frequency of the warnings about being shot at, wasn't too bad. I've presented everything in imperial measurements, even though the standard in the region is metric.

This was a hard journey and took a significant mental and physical toll on all of us who made it. There were a lot of reasons for this: the military attention, the instability of many of the areas we went through, the heat, the gastroenteritis. It became overwhelming to see the layers of trauma and tragedy that Iraqis have suffered, day after day after day. I tried to give an indication of this, but not to dwell too much on the impact these things had on our team. As one early reader said, 'Maybe best to go easy on the breakdowns.' It's true that the best travel writing has a searing honesty at its heart, especially in regards to its protagonists, and I hope you can get a sense of the constant stress that the team was under for much of this journey. But too much self-reflection felt like it would take away from the other stories on the river, which, ultimately, were more important to me. The reason I mention the difficulty here is that, a year and a half later, we're all still recovering from the experience to some degree. None of us regret making the journey, but I'm not sure we'd do it again.

I wrote this book between summer 2021 and autumn 2022, mostly at my desk in Erbil. It was a brutal summer in 2022 – regularly over 50°C, and I lost count of how many dust storms we had. For five or six months it seemed like there was at least one a week. Anything we have had in Erbil was, of course, much more severe in the south.

The situation, all across Iraq, has got worse since we made our journey. I'd always hoped that I'd be able to pen a nice note at the end of the book that would make you feel better about everything, but I can't (without lying to you). There are a few footnotes throughout that suggest the deterioration. Temperatures increased, water was more scarce than ever, and pastoralists in the south sold livestock and moved to cities in greater numbers. In the marshes thousands of buffalo died, and large swathes of the Hawizeh and Central Marshes dried up.

In search of some hope, I spoke again to Azzam Alwash. He was in a philosophical mood. 'Iraq is eternal,' he told me. 'The only constant is change, and Iraq has always suffered. But it will always come back. Have faith in nature.' He outlined opportunities that he sees for Iraq to cooperate with Turkey and Iran on water sharing, to push for modern irrigation methods, to change the country's infrastructure and develop thermal and voltaic energy projects.

Solutions and optimism, even tempered optimism, are rare in Iraq. Azzam is very realistic about the failure of politicians to take the necessary actions, but if you are engaged by the story of the Tigris I suggest reading more of his writing and looking at the work of the NGO he founded, Nature Iraq. Similarly, Jassim Al-Asadi in Chubayish is a singular source of information on the marshes.

There are, of course, plenty of books on the long history of Iraq and Mesopotamia. I learned a lot from reading the work of Gwendolyn Leick, Karen Radner and John MacGinnis. George Roux's *Ancient Iraq* and Samuel Noah Kramer's *The Sumerians* were also very helpful. Paul Kriwaczek's recent book *Babylon* was useful too, and it's always fun to read anything by Irving Finkel of the British Museum.

There are hundreds of great Iraqi authors whose work appears in English. Najim Wali, Ahmed Saadawi, Dunya Mikhail and Shadad Al Rawi are some of the better-known names. During the journey I read *The Watermelon Boys* by Ruqaya Izzidien, which I loved. A friend sends me translations of a different Iraqi poet each month, which has been a delight. The Tigris poems I remember best are from Muhammad Mahdi Al-Jawahiri, Badr Shakir Al-Sayyab and Lamia Abbas Amara. The work of Latif Al-Ani, the father of Iraqi photography, is a window into a very different country in the middle of the last century.

In terms of organisations and individuals doing amazing work, Humat Dijlah, of course, are heroes, and to my mind the best hope for a more positive future for Iraq. Nabil Musa and his Waterkeepers team in Kurdistan are the same. The Nahrein Network in the UK support exciting, important projects. Rashad Salim's expeditionary art is wonderful in so many ways.

I am fortunate to have a group of Iraqi friends from all backgrounds who do their best to help me understand their country. My assessments in this book are largely thanks to their tutoring, and any failings are mine alone.

I've noticed that many Iraqis I work with will often say this when we finish: I am sorry if anything was not perfect. It's a lovely, humble quirk, and every time I hear it I feel privileged to live and travel in this country. And so I will borrow it here. I hope you enjoyed this story, and I am sorry if it was not perfect.

Glossary

Akkad   An ancient Mesopotamian city that became the seat of the Akkadian Empire. Its power peaked between the twenty-fourth and twenty-second centuries BC.

Aramaic   A Semitic language dating to roughly the eleventh century BC. In the context of this book, it is relevant for the Syriacs, who speak a dialect called Turoyo and perform liturgy in a dialect of classical Aramaic, and also for the Mandaeans, who speak a dialect called Mandaic.

Arba'een   A Shi'a observance, meaning the fortieth, marking the martyrdom of Imam Hussein. Each year millions of pilgrims walk between the holy shrines in Najaf and Karbala.

Assyria   The Mesopotamian region that began from the city of Assur and developed into an empire which at its height was likely the most powerful the world had ever seen. Today's Assyrians claim lineage to these ancestors.

ayatollah   High-ranking clergy from the Twelver branch of Shi'ism. In this book I reference two Grand Ayatollahs, Ali Al-Sistani and Ali Khamenei. This is a rare and highly significant honorific recognising their religious knowledge and wisdom.

Ayyubid	The dynasty founded by Saladin, which ruled in the twelfth and thirteen centuries.
Ba'ath party	The Arab Socialist Ba'ath Party in Iraq was a branch of the Ba'ath party that originated in Syria, but later split and became the party through which Saddam Hussein came to power.
Babylon	Babylon was the capital of the Babylonian Empire, which existed in some form from 1895 to 539 BC. Today the remains of the city are still visible, close to Hillah in central Iraq.
Beth Nahrain	The Syriac name for Mesopotamia, meaning 'between two rivers'.
Caliph	The figure historically considered the successor of the Prophet Muhammad. During the period 632–1258 AD, the Caliph was the leader of the Islamic states and of all Muslims.
chaikhana	Teahouse.
chubasha	Islands formed by layers of decomposed reeds in the marshes.
cuneiform	One of the world's earliest forms of writing. Scribes used a reed stylus to cut wedge-shaped marks in clay.
Da'esh	The derogatory Arabic acronym for the militant Islamist group ISIS (Islamic State in Iraq and Syria). I have used ISIS and Da'esh interchangeably, because I heard both frequently.
dengbêj	The Kurdish sung-spoken oral storytelling, literally meaning *deng*, the voice, and *bêj*, to tell.
dishdasha	Ankle-length robe worn by men in Iraq, and throughout the wider region.
diwan	Arabic term for a room for guests.

djinn	A genie, ghost or other spirit.
duba	An oversized *kelek* raft, used in Iraq.
fatwa	A legal decree issued by a religious authority.
guffa	A basket-shaped Mesopotamian boat, made of reeds.
haji	A Muslim man who has completed the pilgrimage to Mecca.
haram	Forbidden by Islamic law.
Hashd al-Shaabi	Umbrella group of mostly Shia militias, formed in 2014 after a *fatwa* from Grand Ayatollah Ali Al-Sistani. Also called the Popular Mobilisation Forces (PMF).
Hisbah	ISIS's so-called morality police.
ICTS	Iraqi Counter Terrorism Service.
IED	Improvised Explosive Device.
IDP	Internally Displaced People.
iftar	The evening meal to break the daily fast during Ramadan.
insha'allah	Literally meaning 'God willing' but used freely to mean everything from 'definitely' to 'no way'.
Kataib Sayyid al-Shuhada, KSS	A brigade of the *Hashd al-Shaabi* with close ties to the Iranian Revolutionary Guard Corps (IRGC) and designated as a terrorist organisation by the US.
kelek	A raft, traditionally made of inflated goat skins and oak logs, used as far back as the time of the Assyrians.

Kermanji	Dialect of Kurdish spoken in Turkey, north-east Syria, parts of northern Iran and also areas in northern Iraq.
Mandaeans (Sabeans)	Likely the smallest ethno-religious group in Iraq, with fewer than five thousand adherents left in the country. Guardians of a unique language, culture and religion.
masgouf	Carp cut open longways and grilled over an open fire. The Iraqi national dish.
mashouf	Broad term referring to the long, flat-bottomed canoes used in the Iraqi marshes.
Mir	A historic title for a ruler, used in this context to refer to a Kurdish tribal leader.
mudhif	Traditional large reed-built home or gathering area from the Iraqi marshes.
mukhabarat	Arabic term for intelligence officers.
mukhtar	Mayor, of a village of town.
NES	North and East Syria.
Newroz	Kurdish New Year, celebrating the arrival of spring. This transliteration is specific for the Kurdish celebration – *Nowruz* or *Nawruz* is more common for Iran, Afghanistan and Central Asia.
pasha	High-ranking position in the Ottoman Empire.
Peshmerga	Kurdish military forces. The name means 'those who face death'.
PKK	The Kurdistan Workers' Party. A Kurdish militant organisation calling for greater Kurdish autonomy in Turkey. The conflict between the Turkish state and the PKK continues in south-east Turkey and northern Iraq.
safina	Arabic term for a large boat or ship.

Saraya Salam   Meaning the 'Peace Brigade', Saraya Salam are a militia under the control of Muqtada Al-Sadr. Before 2008 they were known as the *Jaysh Al-Mahdi* – the Mahdi Army.

Seljuk   The Seljuk Empire was a Turkic empire that ruled large swathes of the Middle East in the eleventh and twelfth centuries.

serifa   In the Iraqi marshes, a *serifa* is a smaller, more humble *mudhif.*

shanasheel   The Iraqi name for an ornate wooden balcony, found in homes in Basra and Baghdad. Today the architectural style has been neglected, and is symbolic of other aspects of endangered heritage.

sheikh   A tribal leader.

Sumer/ Sumerian   The earliest civilisation in Mesopotamia, beginning around 4500 BC.

Sunni Triangle   Name ascribed to the region of central Iraq between Baghdad, Ramadi and Tikrit that is predominately populated by Sunni Muslims.

Syriac   Multiple designations, but in this book it mostly refers to the peoples of the Syriac Orthodox Church and their language, classical Syriac, which is a dialect of Aramaic.

Turkmen   A Turkic ethnic group. In Iraq the Turkmen are often descended from Ottoman migrants from the days of the empire, but some claim lineage back to Turkic warriors employed by the Umayyad and Abbasid caliphs in the seventh and eighth centuries.

Umayyad   A dynasty that created an Empire in North Africa and the Middle East, ruling from 661 to 750 AD. Succeeded by the Abbasids.

Yazidi   An ethno-religious group rooted in the mountains of northern Iraq, but with populations across Syria, Turkey and Iran.

YPG   Meaning the 'People's Protection Unit', the YPG are a Syrian Kurdish military that leads the SDF (Syrian Democratic Forces), who are the defence force of North and East Syria.

Zaza   Zaza are generally considered Kurds but trace their lineage to Iran and speak an Indo-European language that bears no relation to Kurdish. The population exists only in the south-west of Turkey.

Acknowledgements

There is no adequate way to share the gratitude I have to Emily for her role in this journey, this book and my life. She bought into the idea wholeheartedly from the outset, crafted it with care and helped bring us all along the river safely. This story is as much hers to tell as it is mine. She shared her notes from the trip, put up with me moodily writing a book for over a year and even read drafts and offered sensitive critiques. I owe her endless love and thanks for this, and so much more. Her website, www.emilygarthwaite.com, is a good place to see some of her work from the Tigris.

I first had the idea for this book in 2019 and mentioned it to Caroline Michel, who I am incredibly lucky to now have as my agent. When others told me the journey was impossible, Caroline said simply that she saw a beautiful book. She backed it, and me, and I am forever grateful.

Mohamed Amersi was the other early adopter of the idea and helped us think through our storytelling. His foundation supported us financially, and Mohamed and his wife Nadia were invaluable advisers throughout. Emily and I value this immensely, and are proud to have worked with them.

The Abraham Path Initiative, with whom I've had a long relationship, also gave us their financial support, and so I am thankful to Anisa Mehdi and Benjamin Barrows for trusting us, and to Neil and Trudie Prior for helping the journey become a reality.

Pippa Hart, in memory of Ray Rathborne, and Nick Harvey made it possible for us to hire Claudio and create a visual documentation of the river – thank you, and also to Mark Stothert for saving the day.

Mohammed Al-Zaidawi was not on this journey in person but influenced every part of it. I'm privileged to have him as a friend, and this book has benefited greatly from his thoughts.

James Gurbutt has been a patient editor, and the whole team at Corsair a delight at helping bring this to life – thanks also to Zoe Gullen and Susan de Soissons.

There are so many people on and related to the river to whom I am indebted. Our wonderful team, of course. Thank you to Bişar, Angel, Salman, Hana and Claudio for their friendship and hard work. The Tigris has bonded us all for life. Thanks especially to Salman's wife, Liqaa, for pushing him to join us when he was worried about taking time away from his work and family. Thanks to everyone who advised on the journey, and offered contacts and ideas. Thank you to those who helped us solve problems, of which there were many. Thanks to the readers of early drafts of this book for your clear-eyed notes. In no particular order, I am grateful to:

Rashad Salim, Hannah Lewis, Azzam Alwash, Pete Schwartzstein, Sam Sweeney, Alastair Humphreys, Rob Lilwall, Charlie Walker, Julian Sayerer, Laween Mohammed, Miran Dizayee, Shane Winser, Sarah Giaziri, Justin Marozzi, Huda Abuzeid, Matthew Teller, Emma Sky, Claire Hajaj, Thair Ali, Zaab Sethna, Jennifer Hattam, Jaafar Jotheri, Louise Sibley, Jo Cantello, Tim Binding, Anna Bachmann, Stephen Hickey, Sam Nicholls, Elinor Lipman, the

Garthwaites, Norah McCarron, James Wyness, Arik Gabbai, Peter Mellgard, Lawk Ghafuri, Hasan Janabi, Vicky Swyer, Sophy Roberts, John MacGinnis, Zainab Mehdi, Mahmut Sansarkan, Margaret Bowling, Khalil Khalaf Al-Jbory, Karim Wasfi, Hussein Faleh, Shara Dillon, Binwar Rizgar, Dareen Mohammed, Ali Jawad Al-Musaferi.

A final thanks to the small handful of people whose names I cannot mention, but who in anonymity did so much across Turkey, Syria and Iraq for us.

Credits

11   'Debate Between Bird and Fish', in Jeremy Black, Graham Cunningham, Eleanor Robson and Gábor Zólyomi (eds), *The Literature of Ancient Sumer* (Oxford University Press, 2004); Ronald Broadhurst (trans.), *The Travels of Ibn Jubayr: A Medieval Journey from Cordoba to Jerusalem* (Jonathan Cape, 1952)

41–2, 48   Zekeriya Kurşun, 'Does the Qatar Map of the Tigris and Euphrates belong to Evliya Çelebi?', *Journal of Ottoman Studies*, 39 (2012)

47–8   Martin van Bruinessen and Hendrik Boeschoten (trans.), *Evliya Çelebi in Diyarbekir: The Relevant Section of The Seyahatname* (Brill, 1988)

49   Faisal Husain: 'In the Bellies of the Marshes: Water and Power in the Countryside of Ottoman Baghdad', *Environmental History* 19:4 (2014) and *Rivers of the Sultan: The Tigris and the Euphrates in the Ottoman Empire* (Oxford University Press, 2021)

67*   'Tale of Saint Mor Augin', from morauginmonastery. wordpress.com

109–10   Yasin M. Alkalesi, *Iraqi Phrasebook: The Essential Language Guide for Contemporary Iraq* (McGraw-Hill, 2004)

119   Al-Muqaddasi (trans. Basil Collins), *The Best Divisions for Knowledge of the Regions* (Ithaca Press, 2001)

124   Alfred S. Bradford, *With Arrow, Sword, and Spear: A History of Warfare in the Ancient World* (Westport: Praeger, 2001)

126, 246   Gwendolyn Leick, *Mesopotamia: The Invention of the City* (Allen Lane, 2021)

190–1   Karen Radner, *Ancient Assyria: A Very Short Introduction* (Oxford University Press, 2015)

229   Mohammed Mahdi-al Jawahiri, 1962, translated by

Hussein Hadawi and cited in Venetia Porte[
into Art: Artists of the Middle East (British M
Press, 2006); Samuel Noah Kramer, *The Su*
Their History, Culture and Character (Unive
Chicago Press, 1962)

233, 261 Tim Mackintosh-Smith (ed.), *The Travels*
Battutah (Picador, 2002)

290 Wilfred Thesiger, *The Marsh Arabs* (
Penguin, 2007)

304, 305 A. Clement, 'Excursion dans le Kourdistan ottom[
meridional de Kerkout a Ravandouz', Memoire[
de la Societe de Georgraphie de Geneve, 5 (1866).
Quoted in D. T. Potts, '"Un coup terrible de la
fortune", A. Clement and the Qrna disaster of 1855',
in I. J. Finkel and St J. Simpson (eds), *In Context: The
Reade Festschrift* (Archaeopress, 2020)

Emily photographed the entire journey – see www.emily garthwaite.com for more of her work from the river.